Quantum Physics

In Plain English for Absolute Beginners

A Complete, Math-Free Guide to the Rules That Govern Reality From Particles to Technology to the Nature of Existence Itself

C. Louis-Charles, PhD

Cybersoft Publishing LLC

Cybersoft Publishing LLC

Fort Washington, MD 20744

Drclaude.net

First Edition March 2026

Introduction

Why This Book Exists

Start with a kitchen. Right now, someone is making coffee. The kettle heats the water, steam rises, and nothing about the scene feels mysterious. The physics here is the same physics Isaac Newton would have recognized. Objects rest on surfaces. Heat rises. Things fall. We have been living within classical physics our whole lives, and it feels obvious because it aligns perfectly with our instincts.

Now zoom in. Zoom past the water molecules, past the atoms, past the electrons orbiting their nuclei. Down at the scale where the universe does its deepest work, the rules change completely. Particles do not simply sit in one place. They exist in clouds of possibility. A single electron can take two paths at the same time. Two particles separated by vast distances can be so deeply connected that touching one instantly affects the other. These are not science fiction plots. They are the confirmed, tested results of the most successful scientific theory ever written.

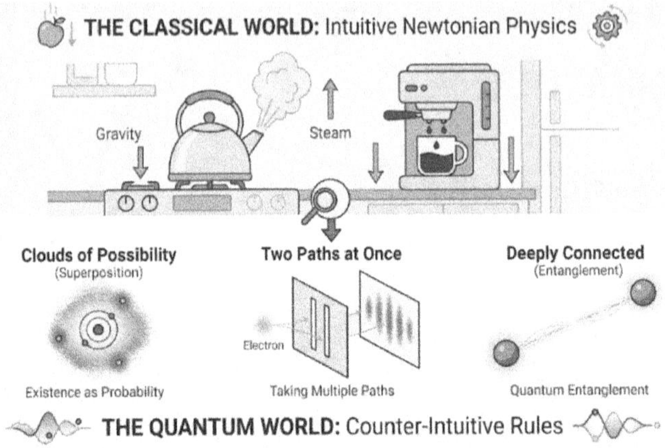

This book is for every person who has heard the word "quantum" and felt a small, quiet panic, as if a door labeled "smart people only" had just swung shut. That door was never locked. The ideas behind quantum physics are genuinely strange, but strange is not the same as impossible to understand. All you need is curiosity and the willingness to let go of a few instincts that work perfectly in kitchens but break down at atomic scales.

What Quantum Physics Actually Is

Quantum physics is the science of the very small. It is the set of rules governing atoms, electrons, and photons (the tiny packets of light). Scientists developed it over roughly 50 years, starting around 1900, after repeatedly encountering experimental results that made no sense under the old rules. Each result was a clue pointing to the same uncomfortable conclusion: the universe at its smallest scale does not behave like a tiny version of the world we can see and touch.

The world you can see and touch follows classical physics. A ball thrown across a room follows a predictable arc. Light from a lamp spreads in smooth waves. This framework works magnificently for everything from building a bridge to landing a spacecraft on the moon. But when scientists probed the very small, classical physics gave wrong answers. Not slightly wrong. Catastrophically wrong. A new framework had to be built from scratch. That framework is quantum physics.

Diagram I.1 - The Two Worlds We Live In

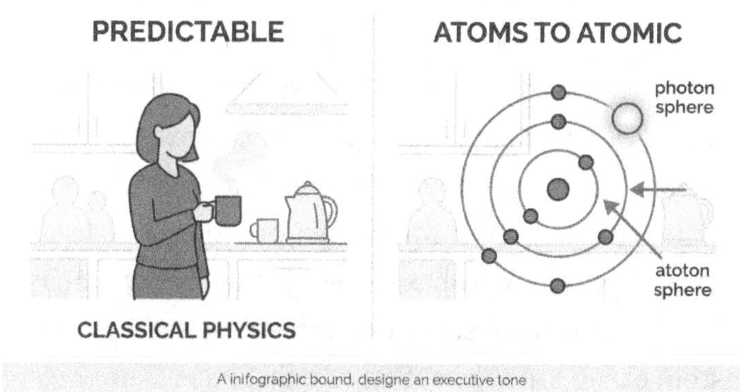

Quantum physics does not replace classical physics. It sits underneath it. For objects larger than atoms, quantum rules average out, and classical rules emerge. Your kettle obeys Newton. The electrons inside the metal of your kettle obey quantum mechanics. Both are true simultaneously, at different scales. The quantum world is not an alternative universe. It is the foundation of the one you already live in, and this book makes that foundation feel ordinary.

The Journey Ahead

This book takes you through 10 chapters, each built around a single big idea. Chapter One asks what went wrong with classical physics in the late 1800s, when confident scientists discovered their framework could not explain the glow of a hot stove. Chapter Two shows that energy does not flow like water. It arrives in little packets, like raindrops. Chapter Three reveals that light is both a wave and a particle at the same time, not as a riddle but as a genuine physical fact.

Chapter Four introduces the uncertainty principle: there is a built-in fog in the universe, baked into reality, not a flaw in our instruments. Chapter Five explores superposition, the mind-bending idea that a particle can exist in two states at once before anyone looks at it. Chapter Six pulls back the curtain on the wave function, the shape that describes where a particle probably is before we measure it. Chapter Seven covers entanglement, the phenomenon Einstein called "spooky action at a distance."

Chapter Eight reveals that particles can pass through solid walls by tunneling through them, and that without this trick, the sun would not shine. Chapter Nine brings everything home: lasers, MRI machines, transistors, and solar cells all depend on quantum physics. The strange science of the very small is not an abstract curiosity. It is the engine of the modern world. Chapter Ten, the final stop, is honest about what physicists still do not know, because the greatest questions in physics remain open debates.

Diagram I.2 - The Book's Journey: Ten Steps on the Quantum Path

Meet the People in This Book

This book has six recurring companions, and none of them are physicists. Maya is a software project manager in her mid-thirties who runs on coffee and relies on her phone for everything. Theo is a recently retired high school history teacher who gardens and wants to stay sharp enough to answer his grandchild's questions. Aisha is an ER nurse who works night shifts and sees the inside of an MRI machine almost every week. Carlos owns a small auto repair shop and has discovered that fixing modern hybrid cars requires dealing with technology rooted in quantum mechanics. Priya is a college sophomore studying graphic design who carries a glowing OLED phone everywhere and is curious about the science she never quite trusted herself to understand. Frank is a retired postal worker who fishes on weekends and has a gift for asking exactly the question you were hoping nobody would ask. Together, they are the reader.

How to Read This Book

Each chapter opens with one of our six friends in an everyday moment: a morning commute, a late hospital shift, a quiet afternoon in the garden. That moment connects to the chapter's big idea. Then the chapter unpacks that idea step by step in plain English, with no equations, formulas, or prerequisites. If you can read a newspaper, you can read this book.

The chapters were designed to build on one another, so reading them in order will give you the fullest picture. But each chapter also stands on its own. If you want to jump straight to entanglement or quantum computing, go ahead. Some ideas in this book will feel uncomfortable at first. That is not a sign you are failing to understand. It is a sign you are succeeding. Quantum physics is genuinely strange, and the discomfort you feel is the same discomfort physicists felt a century ago. You get to skip the decades of confusion.

A Note Before We Begin

Physics has a reputation. It lives on whiteboards covered in symbols, taught in a language that takes years to learn. That reputation has kept too many curious people from ideas they would love. The concepts in quantum physics are not locked behind the math. Math is a precise tool for making predictions, and physicists need it. But the ideas underneath, the actual pictures of how the universe behaves, can be shared in plain language. Physicists do this all the time when they explain their work to family over dinner. That is what this book does.

You do not need to have loved physics in school. You do not need to remember your high school science classes. You need one thing: curiosity. The universe spent about fourteen billion years arranging circumstances that allow you to ask questions about it. It seems only fair to spend a few hours finding out what it has been doing down at the bottom.

The kettle is still on. The coffee is almost ready. Turn the page, and let's find out what's really going on inside that mug.

1 The World That Broke: Why Classical Physics Ran Out of Answers

Opening Scenario

Frank has a system. Every morning, he settles into the old wooden rocker on his porch, pours coffee from a thermos, and opens the newspaper. He reads sports first, then local news, then the science pages if the mood takes him. This particular Tuesday, the mood takes him. A short article explains that researchers discovered an unusual behavior of electrons in a certain type of crystal. Frank reads the paragraph twice, leans back in his fishing hat, and stares at the oak tree in the yard.

Frank spent thirty-one years as a postal worker. He knows how things move. A package follows a path. You toss a ball, and it arcs through the air exactly as your eye expects. The sun rises in the east, climbs, and sets in the west. These are things a person can rely on. His whole working life, the world behaved the way common sense said it should. So when a newspaper article tells him that tiny particles do not obey those same rules, something in Frank resists. It is not stubbornness. It is the perfectly reasonable suspicion of a person who has watched the world work correctly every day for seven decades.

That question, why would physics suddenly stop making sense, is one of the great questions in the history of science. The answer is not that physics stopped making sense. The answer is that common sense, useful as it is for packages, baseballs, and oak trees, is a guide built for a world of large things. When scientists began examining the world inside atoms, they found the guide no longer applied. The rules were not broken. They were simply incomplete. This chapter is the story of how that incompleteness was discovered, what it cost, and what it gave back.

Diagram 1.1 - Frank on the Porch

FRANK'S MORNING READ: A LIGHTLY STYLIZED ILLUSTRATION

FRANK, 70, ENJOYS HIS NEWSPAPER AND THE CALM MORNING LIGHT FROM HIS ROCKING CHAIR.

Why It Matters

You might wonder why any of this concerns you. You are not a physicist. But almost every piece of technology you touch today depends on the discoveries this chapter describes. The phone in your pocket. The solar panel on your neighbor's roof. The laser scanner at the grocery store. The MRI machine at Aisha's hospital. All of them work because scientists, roughly a hundred years ago, discovered that the universe has rules beneath the rules, a deeper layer of behavior that common sense never prepared us for.

This chapter will walk you through the actual chain of evidence. You will see what scientists observed, what they expected to observe, and why the two did not match. You will see what they had to invent to close the gap. By the end, you will understand why quantum physics had to exist. Dreamers did not invent it. It was forced on physicists by data that refused to behave.

There is something personally useful in this story. The scientists who built classical physics were among the greatest thinkers humanity has ever produced. Their framework was spectacularly successful. It still is, for most of the world you can see. Learning where that framework succeeded and where it hit a wall is one of the most honest things a curious person can do. Trust your best understanding, stay alert to its edges, and be willing to rebuild when the evidence demands it.

This chapter covers three things. First, the great triumph of classical physics and why it deserved its reputation. Second, the experiments that fractured that reputation. Third, the first tentative fix planted the seed of everything quantum mechanics would become. When you finish, you will understand why a metal glows red, then orange, then white when you heat it, and why that simple everyday fact turned out to be one of the most disruptive observations in the history of science.

The World Newton Built

To understand why the crisis mattered, you first need to feel the weight of what it threatened. Classical physics, built on the foundations laid by Isaac Newton in the seventeenth century, was not merely a collection of equations in notebooks. It was a worldview. It said that the universe operates like an enormous, perfectly constructed machine. Every part has a position, a velocity, and forces acting on it that can be measured and calculated. If you know the starting conditions precisely enough, you can predict what happens next. Not approximately. Exactly.

This was backed by centuries of evidence so precise it bordered on the miraculous. When astronomers used Newton's laws to calculate where a planet would be in the sky ten years in the future, they were right. Not close. Right. When military engineers used the same mathematics to calculate the arc of a cannonball, they could hit a target miles away. When ship navigators calculated the moon's pull on tides, they predicted tidal heights months in advance with an accuracy that saved lives.

Consider how extraordinary that is. A mathematician sitting at a desk in the 1800s could tell you where the planet Neptune would be in the sky on a specific date decades later. He could do this before Neptune had even been observed. The planet was discovered precisely where the math said it should be. That is not luck. That is a framework that is genuinely and deeply right about something fundamental, like the universe.

Diagram 1.2 - The Classical Clockwork Universe

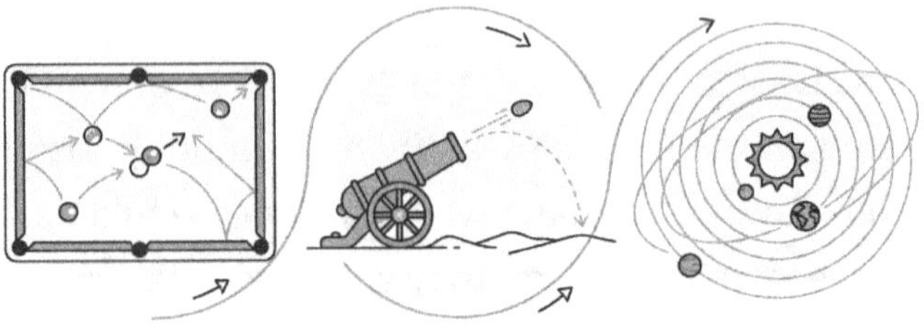

The principles underneath can be stated. Objects have definite properties at every moment. A ball has a precise position, speed, and direction. Forces cause changes in those properties in a smooth, continuous way. And if you know everything about the current state of a system, you can calculate its entire future. There are no surprises. There is no randomness. The universe, in this picture, is a book already written. Physics is just reading it.

This worldview was called determinism. Every event is the inevitable consequence of the events that came before it. A billiard ball travels across the table because it was struck at a particular angle with a particular force. A planet follows its elliptical path because gravity pulls it inward with a precisely calculable strength. Everything chains back to everything else, cause-and-effect, in a seamless mechanical sequence.

By the late nineteenth century, this framework had swallowed up electricity and magnetism, producing a unified theory that predicted the existence of radio waves before they were detected. It explained why gases behave the way they do. It gave scientists a framework for thermodynamics and for understanding how heat flows. The edifice was so solidly built that some leading physicists at the end of the 1800s were heard to say that physics was essentially finished. There were just a few small loose ends to tie up.

Those loose ends would unravel everything. One of them was the behavior of light and heat. Everyone knew that a hot object glowed. Hold an iron over a flame, and it turns red. Heat it further, and it moves through orange and yellow toward white. Blacksmiths knew this. Glassblowers knew this. Potters checking a kiln knew this. What nobody could explain, using all the magnificent machinery of classical physics, was exactly why the colors changed the way they did.

Diagram 1.3 - Planets in Orbit

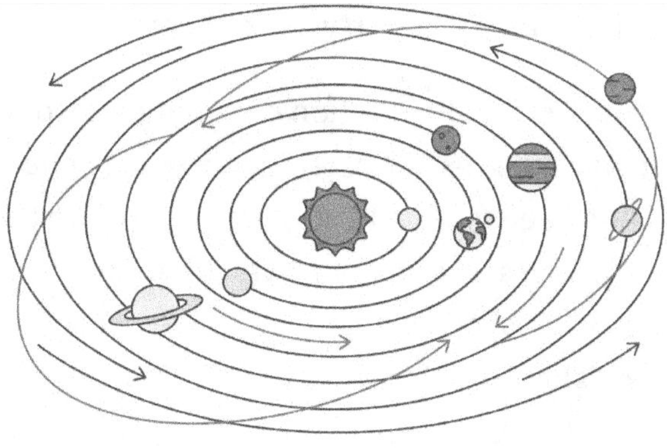

The Strange Glow of Hot Things

Every object, at every temperature above absolute zero, emits radiation. You are doing it right now. Your body is warm, and warm things radiate energy as waves. The waves your body emits are infrared, too long to see with the human eye. When an object gets hotter, those waves shift. They get shorter and more energetic. Eventually, they fall within the visible light range, and the object begins to glow.

Scientists gave a name to the idealized version of this glowing object: a blackbody. A blackbody is a theoretical perfect absorber and emitter. It absorbs every wave that hits it and radiates energy solely as a function of its temperature. Nothing about its material composition or shape affects the radiation it gives off. Just temperature. This made it an ideal testing ground. If you could predict the exact spectrum of light coming out of a blackbody at a given temperature, you would have proven that your theory of heat and radiation was complete.

The spectrum is the distribution of radiation across different wavelengths, a picture of how much energy the object emits at each wavelength. On the left are the long, low-energy waves, such as radio waves and infrared. On the right are the short, high-energy waves, such as ultraviolet light. A blackbody at room temperature radiates most of its energy in the infrared. Raise the temperature, and the chart's peak shifts toward visible light. The blacksmiths and glassblowers were watching this shift happen every time they checked the temperature of their work by eye.

Diagram 1.4 - The Blackbody Glow

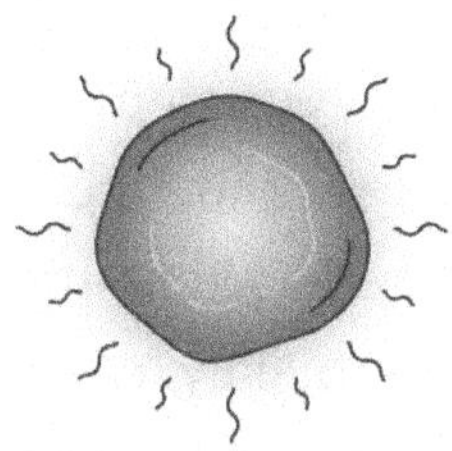

By the 1890s, experimental physicists had measured blackbody radiation with great precision. They had the data. They knew exactly what the spectrum looked like at various temperatures. The challenge was for theorists to explain it from first principles, using only the known laws of physics. This should have been routine. The known laws were the triumphant framework that had spent two centuries predicting everything else in the universe.

The problem started with a classical calculation. Classical physics said that the energy inside a heated box should distribute itself equally among all the possible modes of vibration. Sounds reasonable. Energy spreads out. Spread it equally, and you get a predictable result. Physicists used this rule to try to predict how a blackbody would radiate.

They counted the number of possible modes at each wavelength. And here is where the trouble began.

Shorter wavelengths can fit more standing-wave patterns into a given space. The number of possible modes increases without limit as the wavelengths get shorter. If energy is distributed equally among all modes, and if there are infinitely many short-wavelength modes, the calculation says the object should pour out infinite energy at the short end of the spectrum. Infinite. Not a very large amount. Infinite. That is plainly, obviously wrong. The room you are sitting in is not flooding you with infinite radiation.

This mismatch was so stark that it earned a name: the ultraviolet catastrophe. The classical prediction drew a curve that shot upward toward infinity. The experimental curve did the opposite. It peaked, then fell gracefully back toward zero. The two curves agreed at long wavelengths but diverged violently at short ones.

The Ultraviolet Catastrophe

The physicist who stared longest at this problem was a German named Max Planck. He was careful, conservative, and deeply respectful of the classical framework. He did not want to overturn physics. He wanted to explain a curve on a chart. He spent years trying every available tool of classical theory, only to fail. Then, in 1900, in what he later called an act of desperation, he tried something that had no logical justification in classical physics. He assumed, purely as a mathematical trick, that the energy in the vibrating atoms of a hot object could not take any value it liked. It could only come in specific chunks.

In classical physics, energy is smooth and continuous. You can have any number of units: 3, 3.7, 3.141592. The number can be anything. Planck's trick was to say: What if that is wrong? What if energy comes in discrete, minimum-size packages? What if the universe does not permit the values in between? He introduced a tiny constant, now called Planck's constant, that set the minimum size of these packages. When he plugged this assumption into his calculation, the math worked. The curve came out exactly right. It matched the experimental data perfectly.

Planck was not entirely happy about this. He hoped his constant was just a mathematical convenience, not a statement about physical reality. He thought someone would explain, using normal continuous physics, why the trick worked. That person never came. The trick turned out to be true. Energy really does come in discrete packets. The universe really does have a minimum energy chunk size at any given frequency.

Diagram 1.5 - The Ultraviolet Catastrophe

A tale of two curves: The Ultraviolet Catastrophe.

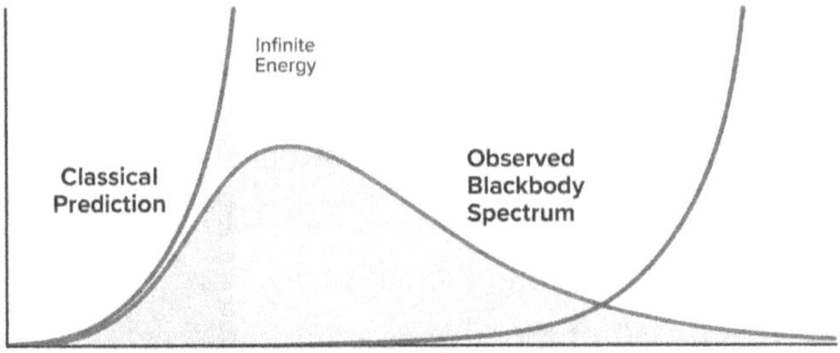

Quantum theory resolves the dramatic divergence at short wavelengths.

To understand why discrete energy packets solve the problem, think about what they do to the number of available modes. The classical calculation went wrong because there are infinitely many short-wavelength modes, and if energy is distributed equally among all of them, you get infinite radiation. Planck's packets change this. A short-wavelength mode requires a minimum amount of energy to activate. If the temperature does not supply enough energy for even one packet, that mode stays empty. The infinity is cut off before it can start.

Think of a vending machine. Classical physics says you can buy anything for any price, as small as you like. Everything participates. Planck's fix says the machine has a minimum price. Some items at the back cost a full dollar. If your pockets only have loose change, those items stay where they are. The high-energy modes at the short-wavelength end are like those expensive items. When the temperature is not high enough to supply a full energy packet, those modes stay quiet. The catastrophe disappears.

Planck was not just adding a fudge factor. He was saying that nature, at a fundamental level, is grainy. Energy does not flow in a smooth, continuous stream. It comes in lumps. The size of the lumps depends on the vibration frequency. This was a crack in the foundation of classical physics. Planck hoped it would seal itself. Instead, someone found that the crack was not a crack at all. It was the first outline of a completely new building. That someone was Albert Einstein, five years later, looking at a completely different problem.

Diagram 1.6 - Planck's Energy Packets

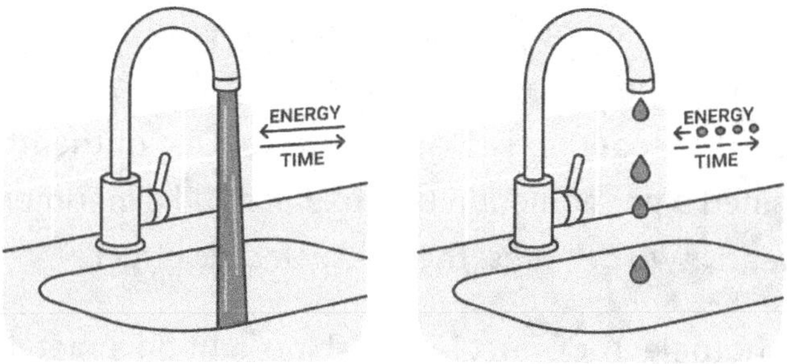

The failure of classical physics to explain blackbody radiation was not seen as a crisis in 1899. It was seen as a puzzle. No one imagined that solving it would require abandoning one of the deepest assumptions of the entire framework: that energy can take any continuous value it likes. That assumption was so deeply woven into the mathematics of physics that questioning it would have seemed like questioning whether numbers could be added together.

That is what makes history so interesting. The revolution did not begin with a grand philosophical rejection of the old order. It began with a patient man trying to match a curve on a chart. The deepest ideas in science often arrive that way. Not as triumphs dressed in banners, but as quiet, reluctant admissions that something small and specific does not add up.

Einstein's Surprising Fix

In 1905, a twenty-six-year-old Albert Einstein was working as a patent clerk in Bern, Switzerland. He was not yet famous. He had not yet published his theory of relativity. But in that one extraordinary year, he published four papers that together reshaped physics. One of them explained a phenomenon that had puzzled experimenters for nearly two decades: the photoelectric effect.

The photoelectric effect is this: shine light on a metal surface and electrons pop out. This is not surprising on its own. Light carries energy, metal contains electrons, and energy can knock electrons free. The surprising part was how the effect behaved when you changed the light. Classical physics predicted a smooth relationship. More light means more energy delivered to the surface, which means more electrons are ejected at higher speeds. Brighter light should knock out electrons more quickly.

Simple, intuitive, completely in keeping with the idea that light is a continuous wave.

The experiments said otherwise. Changing the brightness changed the number of electrons knocked out, but not their speed. Whether the light was dim or blinding, the electrons that escaped always had the same energy, determined entirely by the color, not the brightness. Shine blue light on the metal, and electrons pop out. Switch to red light, and, below a certain threshold, no electrons come out at all, no matter how bright or how long. The behavior did not match the continuous-wave model at all.

Einstein's explanation was radical. He proposed that light itself was not a continuous wave but was made of discrete packets of energy. He called them quanta, though they would later be named photons. Each photon carried a specific amount of energy determined by its frequency and its color. A blue photon carried more energy than a red photon. When a photon struck an electron in the metal, it gave all of its energy to that electron at once. The electron either got enough energy to escape or it did not. There was no accumulation, no gradual build-up. Each interaction was an all-or-nothing event.

Diagram 1.7 - The Photoelectric Effect

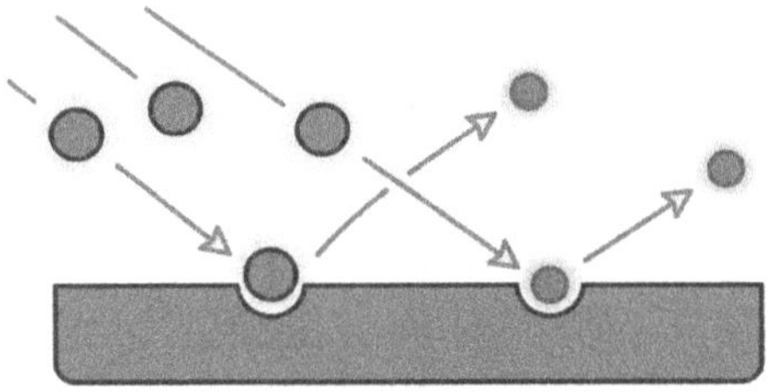

This immediately explains the puzzling results. A red photon does not carry enough energy to kick an electron free from this particular metal. It does not matter how many red photons you throw at it. Each one arrives with the same insufficient energy. No electron escapes. A blue photon, however, carries more than enough. Each one sends an electron flying with a specific energy. Shine more blue light and more photons arrive, so more electrons are kicked out. But each electron still gets the same energy from its individual photon. Brightness controls the number. Color controls the energy. The experiments fit the theory perfectly.

Einstein's photon idea took Planck's mathematical trick and turned it into a statement about reality. Planck had said energy is emitted in packets, but thought that was about the structure of vibrating atoms, not about light itself. Einstein said light was actually made of particles. Or rather, it behaved like a particle when it interacted with matter, even as it also behaved like a wave in other circumstances.

The raindrop metaphor is useful here. A garden hose delivers water in a smooth, continuous stream. Rain is different. Rain comes in drops. Each drop is a discrete parcel. A spot on the surface either gets hit, or it does not. There is no half-drop. Light, in Einstein's picture, works like rain. Each photon is a drop. It either hits an electron with its full energy or misses entirely.

Diagram 1.8 - Raindrops vs. Water Stream

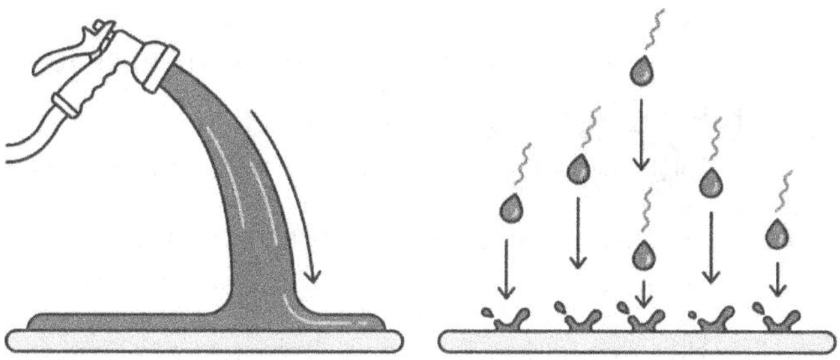

For his explanation of the photoelectric effect, Einstein received the Nobel Prize in Physics in 1921. Not for relativity, which most people assume. For photons. It was not just a clever explanation of one laboratory oddity. It was confirmation that the graininess Planck had glimpsed in blackbody radiation was real and fundamental. The universe really did have a minimum energy chunk size. That single idea would crack open an entirely new physics.

Why Solar Panels Exist Because of This

Frank, sipping his coffee on the porch, might notice the solar panels on the house across the street and think of them as a modern invention, a product of twenty-first-century engineering. In a practical sense, he would be right. But the physical principle that makes a solar panel work is exactly the photoelectric effect that Einstein explained in 1905. Solar panels exist because light arrives in packets.

Here is how it works, in plain terms. A solar panel is made of a material, usually silicon, whose electrons are loosely bound within its crystal structure. When sunlight hits the panel, photons strike its surface. Each photon, if it carries enough energy, knocks an electron free from its position in the crystal. That freed electron moves through the material under the influence of an internal electric field, and that movement of electrons is electric current. Capture enough of those electrons, and you have electricity. The sun's energy has been converted into usable power by exactly the process Einstein described in 1905.

The keyword is enough. Not every photon has enough energy to free an electron in silicon. Infrared photons, the long-wavelength, low-energy ones, arrive but cannot free any electrons. They warm the panel slightly. Visible light photons have enough energy to knock electrons free and generate current. Ultraviolet photons have more than enough energy and do the job, though some of that extra energy escapes as heat. The quantum rule governs the whole operation: each photon either meets the threshold or it does not. There is no partial credit.

Diagram 1.9 - Solar Panels and the Photoelectric Effect

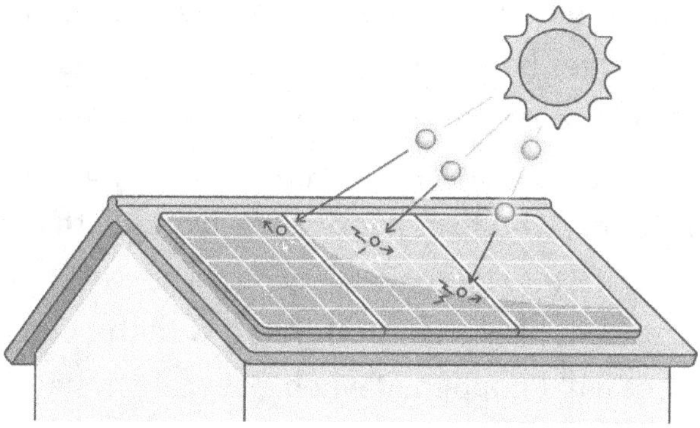

This connection between a foundational physics experiment and a technology that millions of people have on their rooftops is not a coincidence. It is what happens when science gets something genuinely right. The photoelectric effect is not an approximation. It is how the universe actually works at the level of light and electrons. Because it is genuinely true, the engineering built on top of it is reliable across every solar panel in every country.

In 1905, nobody was trying to build a solar panel. Einstein was trying to explain why a laboratory experiment behaved in a way that did not match the theory. He was doing basic research, following the data without a specific application in mind. The application, the technology that converts sunlight to electricity for homes and satellites, came much later, built on the foundation that basic research laid down. This is one of the strongest arguments for funding curious people who want to understand how the world works, even when nobody can yet see why it matters.

Priya, who studies graphic design and takes the train to campus, charges her laptop and phone at outlets whose electricity may come from solar installations whose design traces back to Einstein, who was staring at a puzzling result in 1905. Carlos, who fixes hybrid cars, works with electric motors and battery systems every day. The world that Frank sees from his porch is saturated with the practical offspring of what happened when classical physics ran out of answers.

What the Failure Taught Us

It would be easy to come away from this story thinking that classical physics failed. It did not. It was revealed to be incomplete. Those are different things. A map that correctly shows every road in your county is not wrong when it fails to show the alleyways in a city three states away. It is a map of what it was drawn to show. Classical physics is a magnificent map of the world of large, slow-moving things. It still works. It will always work for baseballs, planets, cannonballs, and tides.

What the failures of the 1890s and early 1900s taught scientists was that the map had edges. There were territories it had never been drawn for, territories of atoms and light quanta and electrons, where the rules were genuinely different. Not just more complicated versions of the familiar rules, but genuinely, fundamentally different in structure. The discovery of those edges and the gradual mapping of the territory beyond them are the rest of the story this book tells.

Scientists had assumed, often without realizing it, that the rules governing large objects would also govern small objects at smaller scales. A billiard ball follows Newton's laws. A dust particle follows Newton's laws. An atom is just a very small particle, so surely it follows the same rules. The answer turned out to be that the analogy is false. The atom is not a small billiard ball. It is a fundamentally different kind of thing, with its own rules and its own logic.

Diagram 1.10 - Classical vs. Quantum Territory

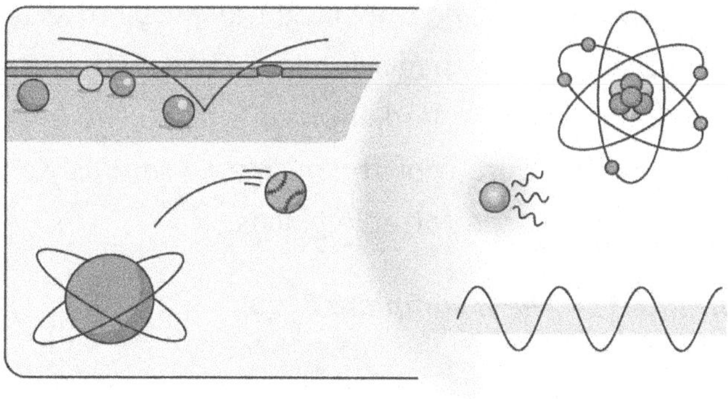

This is one of the most important lessons in science: the rules that work in one regime may not apply in another. Newton's laws work beautifully for everyday speeds. At speeds approaching the speed of light, those laws need correction. At the scale of atoms and smaller, an entirely different framework is needed. None of this makes Newton wrong. A civil engineer designing a bridge does not need general relativity. A baseball pitcher does not need quantum mechanics.

For a baseball, the quantum effects fall far below anything measurable. But for an electron, for a photon of light, for the atoms inside a silicon solar panel or a semiconductor chip or an MRI machine, the quantum effects are not small. They are everything.

Planck's act of desperation and Einstein's bold extension of it also taught scientists something about how discovery works. Progress does not always come from building on what you know. Sometimes it comes from noticing where your knowledge is failing and being willing to ask why. The physicists who dismissed the blackbody problem as a minor puzzle were not bad scientists. They were trusting a framework that had earned enormous trust. The ones who moved things forward took the mismatch seriously enough to follow it into uncomfortable places.

Diagram 1.11 - The Incomplete Map

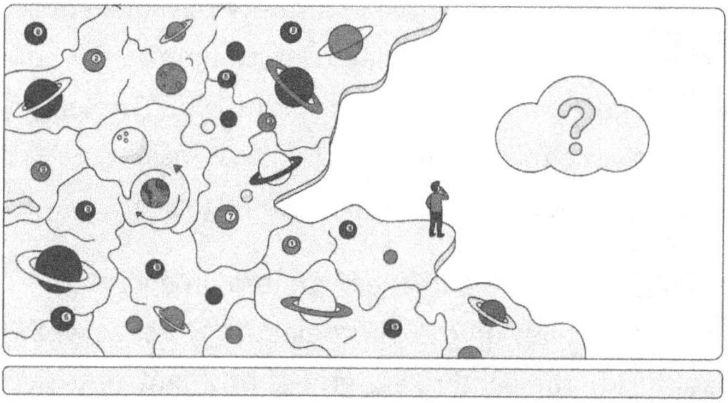

The willingness to follow uncomfortable data is what separates scientific progress from scientific stagnation. When the data and the theory disagree, one of three things is true. Either the data is wrong, or the theory is wrong, or the theory is incomplete. The careful scientist checks the data first, then checks the theory. In the case of blackbody radiation, the data were repeatedly checked by many independent experimenters using increasingly precise instruments. It was right. The theory, magnificent as it was, was incomplete. Accepting that incompleteness opened a door that has never closed.

What This Means For You

You do not need to be a scientist for any of this to matter in your daily life. The discoveries described in this chapter are woven into the fabric of technology you use every single day.

Every time you use a device with a solar charger, whether it is a backpacking lantern, a garden light, or the panels on your roof, you are using the photoelectric effect. The photons from the sun arrive in discrete packets. Those packets hit electrons. The electrons move. You get current. Einstein figured out why that works in 1905. The engineers who built your solar panel relied on that insight, even if they never thought about it in those terms.

Every digital camera and every camera on every smartphone relies on the photoelectric effect as well. The sensor inside the camera is a grid of tiny light-sensitive cells. Each cell collects the electrons freed by incoming photons. More photons, more electrons, brighter pixel. Fewer photons, fewer electrons, darker pixel. The entire image is built from discrete photon-electron interactions, quantum events happening millions of times per second every time you take a photo. The selfie you posted this morning is, at the level of its creation, a quantum mechanical event.

The LED screen you are reading this on also operates through quantum principles. Energy gaps within its semiconductor material determine the colors an LED emits, and those energy gaps are quantum-mechanical in origin. The fact that a blue LED emits blue light rather than green or red light is a consequence of quantized energy levels within the material. Planck's constant, introduced as a desperate mathematical trick in 1900, governs the color of the light on your screen.

Aisha, working a night shift in the emergency room, passes an MRI machine in the hallway. Carlos, diagnosing a hybrid car in his shop, reads data from an electronic control unit built from semiconductor chips. Theo, listening to an audiobook on wireless earbuds, receives a Bluetooth signal generated by quantum-designed electronics. Priya, designing graphics on her laptop, works on an OLED screen whose colors are determined by quantum energy levels. Every one of them lives in a world that would be impossible without the discoveries described in this chapter.

Diagram 1.12 - Quantum Physics in Everyday Life

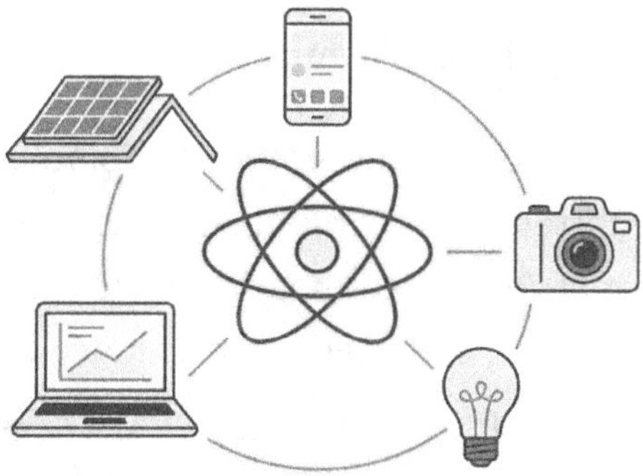

Frank does not need to know any of this to enjoy his morning coffee. But here is what happens when he does know it. The newspaper article about electrons in crystals stops being opaque jargon and starts being a familiar story. He recognizes the characters: the electrons, the energy packets, the threshold effects. He knows how the plot goes, because he has already followed it from the beginning. He has seen where it started, right here in the glow of a heated piece of metal and the puzzled faces of nineteenth-century physicists who expected to find a simple answer and found a revolution instead.

Takeaway

Classical physics was one of the greatest achievements in human intellectual history. For two hundred years, it described the world with astonishing precision. Planets, tides, projectiles, electric currents, magnetic fields: all of them yielded to Newton's framework and the science built on it. The people who built that framework were brilliant, careful, and thorough. They had every reason to believe they were closing in on a complete description of reality.

They were not wrong. They were incomplete. The distinction matters. When scientists at the end of the nineteenth century turned their instruments toward glowing hot objects and tried to predict what they would see, their predictions were catastrophically incorrect. Not a little off. Infinitely wrong. That failure, called the ultraviolet catastrophe, was the crack in the foundation. Max Planck filled the crack with a mathematical trick: the idea that energy comes in discrete packets, not in a smooth continuous flow. He did not believe the trick was real. He thought it was a temporary fix.

Five years later, Albert Einstein took that trick seriously and showed that it was not a trick at all. Light really does arrive in discrete packets, which we now call photons. Each photon carries a fixed amount of energy determined by its color. When a photon strikes an electron, it either has enough energy to knock the electron free, or it does not. There is no partial delivery. That insight explained the photoelectric effect, won Einstein the Nobel Prize, and planted the seed from which the entire field of quantum mechanics would grow.

The practical harvest of these two discoveries is everywhere around you. Solar panels work because photons knock electrons free. Digital cameras work because photons knock electrons free. LED screens work because quantum energy levels determine the color of light they emit. The transistors in every computing device on Earth operate on quantum principles within semiconductor crystals. A quantum world underlies the classical world that Frank watches from his porch, which he cannot see, but that makes all of it possible.

You are now standing at the beginning of that quantum world. You know why it had to be discovered. You know what failure looked like, what data forced the discovery, and what the first tentative answers were. The next chapter takes the story forward: not just that energy comes in packets, but also what it means for light itself to sometimes behave like a wave and sometimes like a particle. That double nature turns out to be one of the most deeply true things we have ever learned about the universe. Frank might need a refill of his coffee before that one. It gets stranger. But strange, as you will see, can also be wonderful.

2 The Packet Problem: How Energy Stopped Being a River and Became a Rain

Opening Scenario

Aisha pushed through the hospital's side door at ten past six in the morning. The sun was coming up behind the parking structure, slanting long orange light across the asphalt. She had been on her feet for twelve hours. The fluorescent lights inside had been humming steadily since she arrived the night before, and now here was this other light, ancient and massive, pouring itself across the city like it had been waiting all night. She stood in it for a second and let it warm the side of her face.

She noticed the heat lamps above the ambulance bay glowing orange against the pale morning sky. She noticed her hybrid car at the charging post, its indicator light blinking blue. She noticed the neon letters of the pharmacy across the street, still lit, their red-orange glow steady and patient. Energy was absolutely everywhere. In the sunlight, hitting her face. In those glowing lamps. In the current flowing into her car's battery. Energy moved through all of it, doing work, making light, keeping things warm.

What Aisha did not yet know, standing in the parking lot after a long night shift, was that all that energy had something surprising in common. It did not flow. Not really. Not the way water flows from a tap in a smooth, unbroken stream. Instead, every bit of it arrived in tiny separate packets, like raindrops. Each packet was a complete, indivisible unit. You could not have half a packet. Energy moved through the universe the way rain falls: one drop at a time, each drop whole, each drop discrete. The idea sounds small. It is not. It is one of the two or three most important ideas in all of science.

Diagram 2.1 - Energy as Rain vs. River

Why It Matters

Most of us learned in school that energy flows. We were told that heat spreads out, that electricity runs through wires, and that light travels in waves. None of that is wrong, exactly. But it is incomplete. It leaves out the most important thing about energy at the atomic level: it does not travel in a continuous stream. It travels in chunks. Fixed-size chunks. And those chunks cannot be broken down any further.

This matters in your daily life in immediate, practical ways. The LED light in Aisha's phone works because of energy packets. The screen on that phone works because of energy packets. The solar cells that power some homes work because of energy packets. The fact that heated metal turns red before it turns orange, then white, is explained by energy packets. Without quantization, none of those things make sense. With it, everything falls into place.

This chapter walks you through how physicists discovered that energy comes in chunks, why that discovery was so hard to accept, and what it means for the world you live in. You will come away with a word worth knowing: quantum. Not as jargon. As a real physical thing. A quantum is the smallest possible chunk of something. You will understand what that means by the end of this chapter, and you will never look at a glowing light bulb quite the same way again.

You will also meet two of the most important ideas in physics: Planck's constant, the number that sets how big each energy packet is, and Einstein's insight that light itself is made of those packets. Together, these ideas launched the quantum revolution. Neither man set out to start a revolution. Both were trying to explain something strange that the data kept insisting on.

Diagram 2.2 - Aisha Under the Fluorescent Light

The Old Picture: Energy as a River

Before we can understand why the quantum idea was such a shock, we need to understand what physicists believed before it. The classical picture of energy is intuitive. It matches what we see every day. And it is, at the atomic scale, fundamentally incomplete.

Picture a garden hose. You turn it on, and water streams out smoothly. If you turn the handle slightly, you get slightly more water. You can adjust the flow to any rate you like. There is no minimum amount. There is no smallest possible unit you have to wait for. The flow is continuous.

Classical physicists imagined energy working the same way. Heat radiating off a hot surface should flow out in a smooth, continuous stream. Light emitted by a burning filament should spread evenly across all possible frequencies. If you want more heat or more light, you turn up the temperature a little and get a little more of both.

The amount of energy coming out at any frequency should increase smoothly and predictably.

For large-scale, everyday situations, this picture works perfectly well. Engineers used it to design steam engines and heating systems. It gave correct answers at the scale of things you can touch and see. The trouble started when physicists tried to apply this smooth-river picture to what was happening inside a glowing solid, when individual atoms were radiating energy.

There was a specific problem with a dramatic name: the ultraviolet catastrophe. If energy really flows continuously, the mathematics says a heated solid should radiate more and more energy at shorter and shorter wavelengths, without any limit. A hot object should pour out infinite energy in the ultraviolet range. Infinitely bright ovens do not exist. But the math of the classical picture produced that result, every single time.

This was not a minor discrepancy. It was a catastrophic failure of the theory. Real hot objects radiate light in a characteristic curve. The output rises, peaks at a certain frequency, and then falls off. It does not explode toward infinity. The classical picture could not explain that shape. Every attempt to derive it using smooth, continuous energy failed. This was the problem that German physicist Max Planck sat down to solve in 1900. What he found when he finally cracked it surprised even him.

Diagram 2.3 - The Ultraviolet Catastrophe

RESOLVING THE ULTRAVIOLET CATASTROPHE: CLASSICAL PREDICTION VS. OBSERVED REALITY

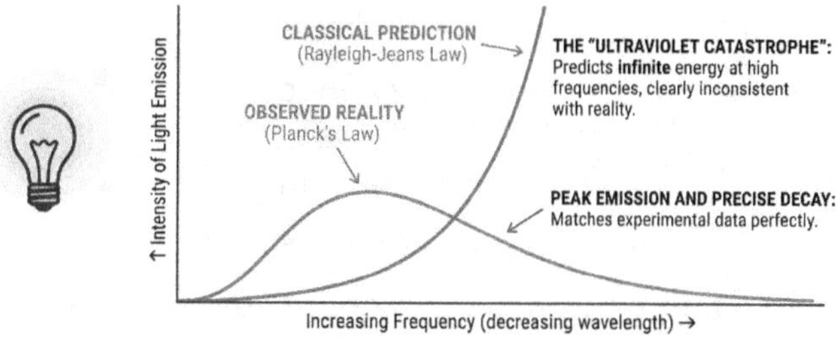

QUANTUM THEORY RESOLVED THE DISCREPANCY, SHOWING THAT ENERGY IS EMITTED IN DISCRETE 'QUANTA'.

Planck's Desperate Idea: The Smallest Possible Chunk of Energy

Max Planck was not a radical thinker. He believed deeply in classical physics and had spent his career working within it. When he set out to solve the radiation problem, he was not trying to overturn anything. He was trying to fix a small crack in a structure he trusted completely.

He worked on it for years. In a desperate act, he tried something he did not really believe in. He assumed that atoms cannot emit energy continuously. Instead, they can only emit energy in specific fixed amounts. Each chunk would be proportional to the light's frequency. Higher frequency, bigger chunk. Lower frequency, smaller chunk. The size of the smallest possible chunk would be set by a fixed number: Planck's constant.

When he plugged this assumption into his equations, the math worked perfectly. The curve matched the experimental data exactly. It was a stunning success. The problem was that Planck did not know what to make of it. He thought the chunk idea was just a mathematical trick. He spent years trying to derive the same result without the chunks, and he never succeeded.

Planck's constant is a very small number. In everyday life, the chunks it describes are completely invisible. When you turn on a lamp, the stream of energy packets is so dense and rapid that it appears to be a continuous flow. The chunkiness only becomes obvious at the atomic scale. This is why the classical, continuous picture works fine for engineering. It is a very good approximation. But at the atomic level, the approximation breaks down.

Think of it this way. Stand a mile from a sandy beach. The shore looks like a smooth, continuous strip of pale color. Walk up close, and you see millions of individual grains. Each grain is a discrete unit. At a distance, the chunks are invisible. Up close, all you can see is the chunkiness. Energy is like that. At human scales, a river. At atomic scales, always a rain of individual grains.

The name Planck chose for his chunks was quanta, the plural of the Latin word quantum, meaning "amount" or "how much". A quantum of energy is a fixed, indivisible amount. It is the smallest unit of energy that can be emitted or absorbed at a given frequency. You cannot have fewer than one quantum. You cannot receive 2.5 quanta. You get whole quanta, and nothing in between.

This is the staircase idea. Imagine energy levels as a staircase rather than a ramp. On a ramp, you can stand at any height. On a staircase, you can only stand on whole steps. You cannot balance between steps. Atoms can only absorb or emit specific fixed amounts of energy, corresponding to the fixed heights of the steps on their staircase.

Planck announced his result in 1900. Almost no one recognized what he had done. The physics community treated it as a clever mathematical technique rather than a revelation about the nature of reality. It would take a young patent clerk in Switzerland, and a few more years, for the true importance of Planck's discovery to become clear.

Diagram 2.4 - The Staircase and the Ramp

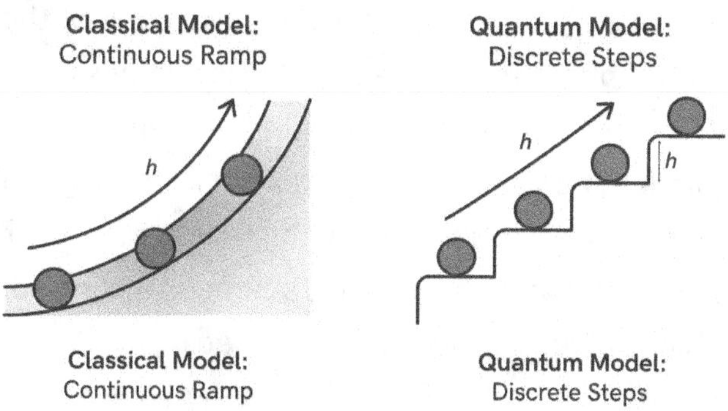

Einstein and the Idea of Photons

Albert Einstein read Planck's work and saw something in it that Planck himself had not fully committed to. Planck had assumed that atoms emit and absorb energy in chunks. But he left open the possibility that light itself, traveling through space, was still a continuous wave. Einstein went further. In 1905, he proposed that light is not emitted in discrete chunks. It travels as chunks. Light is made of particles, each one a discrete packet of energy.

Einstein called these packets light quanta. Later, they were given the name photons, the name we use today. A photon is a single, indivisible particle of light. It carries a fixed amount of energy determined by its frequency. Blue light has a higher frequency than red light, so blue photons carry more energy than red photons. Ultraviolet photons carry even more energy. Infrared photons carry less energy than red ones. The relationship is direct: higher frequency means more energy per photon, lower frequency means less.

This was an extraordinary claim. Light was one of the best understood phenomena in all of physics. James Clerk Maxwell showed in the 1860s that light is an electromagnetic wave, and his equations predicted almost everything about its behavior with great accuracy. The idea that light was also a particle seemed to contradict everything wave theory had established.

Einstein was focused on something very specific: an experiment the wave theory of light could not explain, called the photoelectric effect. What he showed was that the photoelectric effect only makes sense if each electron in a metal is being hit by a single photon, not by a wave.

The electron absorbs the photon in its entirety, and if that photon carries enough energy, the electron gets knocked free. If it did not carry enough energy, nothing happens, no matter how many photons arrive.

This was deeply strange. Energy is not just emitted in chunks. It is also absorbed in chunks. The chunks have to be matched: an electron absorbs exactly one photon at a time, and the photon has to be energetic enough to do the job. You cannot add up the energy of a thousand feeble photons to dislodge an electron. Each photon is a complete transaction.

Einstein published his photon paper in 1905, the same remarkable year as his special theory of relativity. Neither was immediately celebrated. The photon paper was viewed with deep suspicion. It seemed to undermine everything Maxwell had built. Even Planck was skeptical of the photon proposal for years. History vindicated Einstein. In 1921, he received the Nobel Prize in Physics, and the citation specifically recognized the discovery of the law of the photoelectric effect, not relativity. That tells you something about how important physicists judged this contribution to be.

Think about what Aisha was standing in that morning outside the hospital. Sunlight is a stream of photons, billions of which arrive every second. Red, orange, yellow, blue. Each color corresponds to a different frequency, each frequency to a different photon energy. When the photons hit Aisha's skin, they were absorbed. The warmth she felt was billions of tiny energy transactions happening all at once.

Diagram 2.5 - A Photon as a Sphere with a Wave Trail

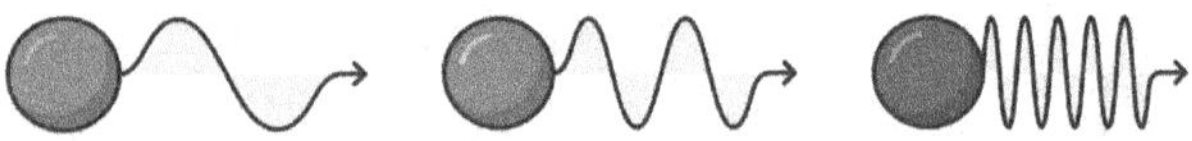

Why Light Has Color: Atoms Only Release Specific Packet Sizes

You have almost certainly noticed that neon signs come in specific colors. Neon itself makes that orange-red glow. Mercury makes blue-white. Sodium makes a deep yellow-orange, the color of old streetlamps. These are not arbitrary. They are not the result of adding a dye or a coating. They come from the atoms themselves, and they reveal something profound about how atoms work.

Here is what is happening inside a neon tube. The tube contains neon gas at low pressure. When you run electricity through it, electrons from the current crash into neon atoms and give them energy. The neon atoms absorb that energy, and their electrons are pushed up to higher energy levels. But those higher levels are unstable. The electrons quickly fall back to their original level, and as they do, they release the absorbed energy as a photon.

The key: the energy difference between the two levels in a neon atom is fixed. It is a specific number. It is the size of one step on Neon's particular staircase. Because that energy difference is fixed, the photon that gets released always has exactly that much energy. Because photon energy and frequency are directly linked, the photon always has exactly that frequency. Because frequency determines color, the photon is always exactly that color. This is why neon signs are always the same color. It is not a coincidence. It is a necessity.

Every element in the periodic table has its own unique set of energy level differences. That means every element has its own unique set of colors it can emit. Hydrogen emits red, blue, and violet. Sodium emits two very close lines in the yellow-orange range. Calcium emits lines in red, orange, and blue-green. Iron emits hundreds of lines spread across the visible spectrum. Each set of lines is as unique to that element as a fingerprint is to a person.

This fingerprint of light is called an emission spectrum. It is one of the most useful tools in all of science. Astronomers looking at distant stars can read off exactly which elements are present just by identifying the pattern of colors. The technique was used to discover helium before it was found on Earth. Astronomers spotted its spectral fingerprint in sunlight in 1868, twenty-seven years before anyone isolated it from a rock.

Bohr's model of the atom, developed by Niels Bohr in 1913, was the first attempt to explain why atoms have discrete energy levels. He proposed that electrons can only occupy specific orbits around the nucleus, each with a fixed energy. When an electron jumps down to a lower orbit, it releases a photon. When it absorbs a photon of just the right energy, it jumps up. Later chapters will show more sophisticated theories. But Bohr got the essential point right: electrons live on a staircase, not a ramp.

Think about fireworks. The brilliant colors come from specific metal salts packed into the shell. Strontium makes red. Barium makes green. Copper makes blue. When the shell explodes, heat excites electrons in those atoms. As the electrons drop back down, they release photons. Each metal emits its characteristic colors. Every burst of color in a fireworks display is a visible demonstration of quantized energy levels.

Aisha works under fluorescent lights all night. Fluorescent tubes contain mercury vapor. Electric current excites mercury atoms, which emit ultraviolet photons. Those photons strike a phosphor coating inside the tube, which re-emits them as visible light. The particular shade of white produced by hospital lighting results from the phosphor blend chosen by the manufacturer. The trigger, though, is always mercury atoms emitting photons at their own specific frequencies, one packet at a time.

Diagram 2.6 - Atom with Electron Jumping Between Shells

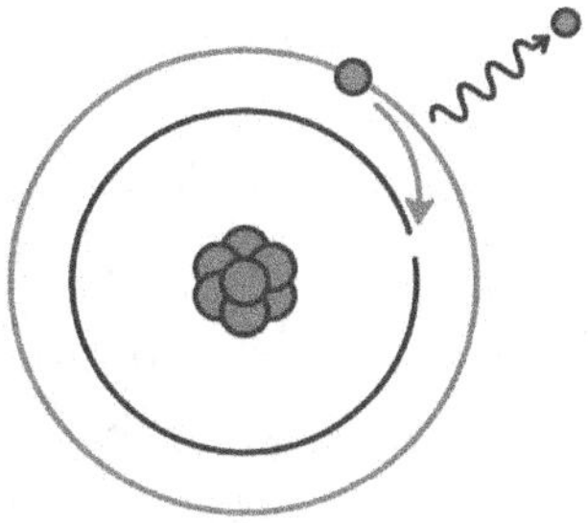

The Rainbow Inside an LED Bulb

When Aisha got home after her shift and turned on a light, it was almost certainly an LED. LED stands for light-emitting diode. Understanding how it works is a direct window into the quantum world.

A diode is a device made from semiconductor materials. Semiconductors sit between conductors (like copper, which lets electrons flow freely) and insulators (like rubber, which stops them). The semiconductor in an LED is designed so that, when electrons flow through it, they move from a higher-energy zone to a lower one. As they make that transition, they release the energy difference as a photon.

The color of the photon depends on the size of the energy gap the electron crosses. Engineers choose semiconductor materials with specific energy gaps to produce photons at specific frequencies, corresponding to specific colors. A small energy gap produces a low-frequency photon, such as red or infrared light. A larger gap produces a higher-frequency photon, such as a green or blue one.

This is why LED technology took so long to produce blue LEDs. Red and green LEDs were available by the 1960s and 1970s. Blue LEDs required semiconductor materials with a much larger energy gap. Finding the right material with the right properties was genuinely hard. The breakthrough came in the early 1990s, work for which three Japanese scientists were awarded the 2014 Nobel Prize in Physics. Blue LEDs made white light possible by enabling the combination of red and green to produce a broad-spectrum white glow.

The LED in Aisha's phone screen is a close cousin of this technology. OLED screens use organic compounds as the light-emitting material, but the principle is the same: electrons transition from a higher to a lower energy level, releasing photons. The compounds in each pixel are chosen to produce red, green, or blue photons. When you look at the screen and see a photograph, you are seeing millions of tiny quantum transitions happening sixty times per second, each one a single electron releasing a single photon of a specific color.

Carlos, who fixes hybrid cars, has probably replaced LED taillights on customer vehicles. The red of a modern LED taillight comes from a semiconductor material whose energy gap is calculated to produce photons in the red range. Before LEDs, red taillights used a white incandescent bulb behind a red filter. The filter just blocked all the non-red photons. Wasteful. An LED taillight only makes red photons in the first place, which is far more efficient. The color of an LED is not painted on. It is engineered into the semiconductor's atomic structure and into the specific size of the energy gap the designer chose. Color, in an LED, is a direct consequence of quantum energy levels.

Diagram 2.7 - LED Bulb with Photons Emerging at One Color

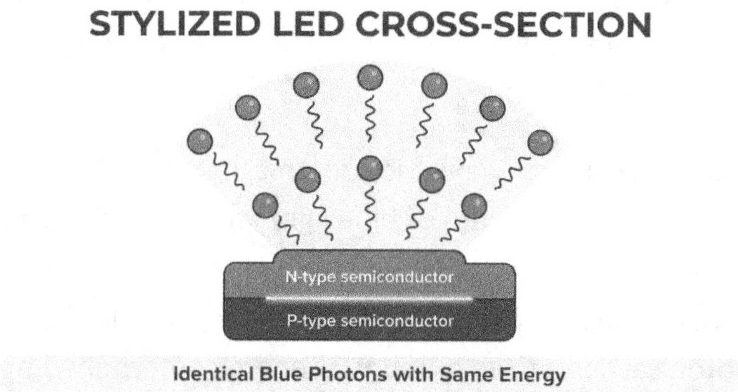

The Photoelectric Effect in Plain English: And Why Your Solar Panel Exists

Here is a puzzle that baffled physicists for years before Einstein solved it. When you shine light on a metal surface, electrons can be knocked loose. This is called the photoelectric effect. It was first properly observed in the late nineteenth century, and it behaved nothing like the classical wave theory of light predicted.

The classical expectation was simple. Light is a wave. Its energy is related to its brightness. So bright light should knock out electrons, and dim light should take longer but eventually accumulate enough energy to do the job. That is not what happens. Whether or not an electron gets knocked loose depends entirely on the frequency of the light, not its brightness. Light above a certain frequency threshold knocks out electrons immediately, even when very dim. Light below that threshold never knocks out electrons, no matter how bright it is. You could flood the metal with low-frequency light all day, and nothing would happen. But the tiniest trickle of high-frequency light, and electrons start flying.

Under the classical wave picture, this makes no sense. Why would a dim beam of blue light knock out electrons when a blindingly bright beam of red light cannot? Einstein's photon picture explains it immediately. Each photon carries a fixed amount of energy determined by its frequency. High-frequency photons carry more energy. Low-frequency photons carry less. To knock an electron loose, you need to give it enough energy to escape. If a single photon carries enough energy, the electron escapes. If it does not, the electron cannot escape, no matter how many photons arrive. The threshold frequency is the frequency at which a single photon carries exactly the minimum energy needed.

This is the quantum transaction: one photon, one electron. If the photon is energetic enough, the transaction completes, and the electron escapes. If the photon is not energetic enough, the transaction fails. Intensity determines how many attempts happen per second, but it does not change the outcome of each attempt. The energy has to arrive in a single, sufficiently large packet.

Solar cells work because of the photoelectric effect. When photons from sunlight hit a solar cell, the energetic ones knock electrons loose. Those free electrons become the electric current. Photons that are not energetic enough pass through or are absorbed as heat.

Only photons with energies above a certain threshold contribute to a solar panel's output. Infrared photons pass through or are converted into heat. Visible and ultraviolet photons are the productive ones. Engineers match the semiconductor material to the photon energies most abundant in sunlight.

Aisha's hybrid car was sitting at the charging station. Some of that power might have come from solar panels. Those panels were running the photoelectric effect billions of times per second, turning photon packets into electrons, electrons into current, and current into stored charge that would eventually move her car. A chain of quantum events, stretching from Planck's chalkboard in 1900 to the parking lot outside the hospital.

The photoelectric effect appears in every digital camera sensor. When photons hit each tiny detector, they knock electrons loose. The number of electrons released indicates how bright the light was at that spot. The result is a digital image. Photography, in the digital age, is the photoelectric effect, recorded in a grid.

Diagram 2.8 - Solar Panel with Photons Knocking Electrons Free

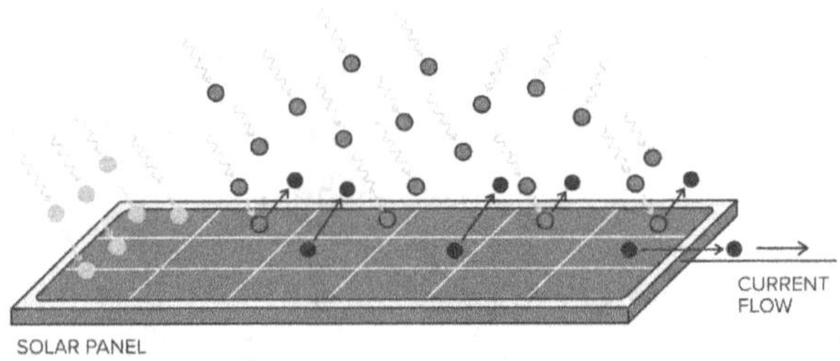

SOLAR PANEL PHOTONES TO FREED EISEED ELECTRONS

Why Heated Metal Changes Color: From Red to Orange to White

There is a practical, everyday mystery that quantization explains beautifully. Aisha walked past a demonstration of it every day: the orange glow of the radiant heat lamps above the ambulance bay. Those lamps get hot and emit light. But why orange? Why not green? And why does metal, when heated even hotter, change from red to orange to yellow to white? The answer is in the packets.

When you heat a solid object, you are giving energy to its atoms. Those atoms vibrate faster and emit photons as they shed energy. A cooler object emits mostly low-energy, low-frequency photons in the infrared range. You feel them as heat but cannot see them. As the object heats up, some atoms gain enough energy to emit visible photons. The first visible photons are the lowest-frequency ones, which are red. So the first sign that something is glowing is a dull red glow. Red photons appear first because red is the lowest-frequency color in the visible spectrum.

Heat the metal more, and higher-frequency photons begin to appear alongside the red ones. Orange. Then yellow. As the temperature climbs further, more and more colors are represented. Eventually, the object emits a significant number of photons across the entire visible spectrum. When you mix all the colors of the visible spectrum in roughly equal amounts, you get white light. This is why a very hot object, like a star's surface or the inside of a furnace at its highest setting, glows white or even blue-white.

This is precisely the behavior that Planck's quantization explains. The shape of the emission curve, the way it peaks at a frequency that shifts with temperature, is exactly what his quantum hypothesis produces. The classical smooth-river picture predicted an explosion toward infinity that never happens. The quantum rainstorm picture produces a curve that rises, peaks, and falls, matching every observation.

The color of a star is a direct readout of its surface temperature. Red stars are cooler. Yellow stars like our sun are medium temperature. Blue-white stars are very hot. Astronomers can determine a star's temperature just by measuring its color, because the relationship between peak emission and temperature is governed by the same quantized radiation that Planck cracked in 1900.

The heat lamps above the ambulance bay are a modest version of this. Tungsten filaments heated to a high temperature glow orange-red. Those atoms emit mostly infrared photons, felt as heat, and some visible red and orange photons, seen as glow. The color of the glow is a thermometer written in the language of photons.

Carlos has seen the color sequence when grinding or welding. A spark starts red, brightens to orange, and flashes white at its hottest. Each stage is at a different temperature, each with a different peak in the photon emission. What looks like a shower of sparks is a live demonstration of Planck's law.

Diagram 2.9 - Glowing Metal Changing Colors

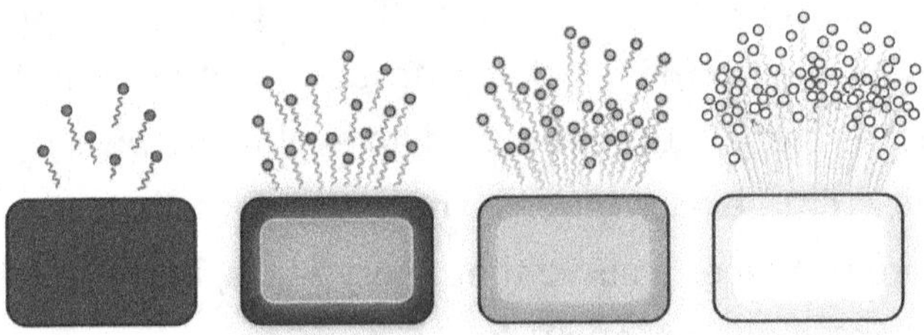

What This Means For You

You interact with quantized energy dozens of times every day, and now you have the vocabulary to see it.

Every light source in your home is a quantum device. If you have incandescent bulbs, the glow comes from a hot filament that emits photons according to a distribution determined by its temperature. The warm orange color is not a design choice. It is the shape of the emission from a tungsten wire at around 2700 Kelvin. If you have LED bulbs, each photon is produced by an engineered quantum transition in a semiconductor. The label on the box that says 2700 K or 5000 K indicates where the peak of the photon emission lies on Planck's curve.

The screen you are reading this on is emitting photons millions of times per second. Each pixel produces photons of red, green, or blue by triggering quantum transitions in organic compounds or semiconductor LEDs. Your eye receives those photons and converts them to electrical signals in your retina. That conversion is itself a quantum event: photons are absorbed by molecules called photopigments, and the absorbed energy triggers a chemical change that sends a signal to your brain. From pixel to retina, the entire process is a chain of quantized events.

The barcode scanner at the grocery store uses a laser. A laser produces photons all of the same frequency. The specific color is determined by the quantum energy level transition used to produce the photons. When the scanner sweeps across a product label, reflected photons carry information that a detector converts to an electrical signal. Laser technology would not exist without the concept of the photon.

Your television remote sends signals using infrared photons with a specific energy determined by the LED inside, which is an engineered quantum transition. The sensor in your TV is sensitive to that specific frequency and ignores others. This is why your TV does not respond to every photon in the room.

Even the color of the sky above Aisha's parking lot that morning is a quantum story. The atmosphere scatters short wavelengths, the blue ones, more than long wavelengths, the red ones. The blue sky you see is composed of photons that air molecules have redirected toward you. The red and orange of the sunrise are photons that traveled a long path through a thick atmosphere, while the blue was scattered away. Every color in nature is a story about which photon energies are being emitted, absorbed, scattered, or transmitted.

Diagram 2.10 - Aisha's Morning: Energy Packets Everywhere

Takeaway

Aisha walked to her car that morning, carrying twelve hours of exhaustion, a hospital badge, and a thermos that was now stone-cold. But she was also wading through a rain of photons that had traveled ninety-three million miles to reach her face. Each one was a complete, indivisible packet of energy. Each one carried exactly as much energy as its frequency required, not a fraction more, not a fraction less.

This is what Max Planck discovered by accident in 1900. He was trying to fix an equation, not change the world. But the fix required a new idea: energy comes in chunks. You cannot emit half a quantum. You cannot absorb a quarter of one. The universe, at its smallest scale, is not a river. It is raining.

Einstein took that idea further in 1905 and showed that light itself is made of those chunks. Each photon is real, discrete, and indivisible. The energy it carries depends on its frequency. That single fact explains why neon signs are always the same color, why solar panels exist, why LEDs can be engineered to emit exactly one color, why a hot iron glows red before orange before white, and why no amount of dim red light will ever knock an electron loose from the right kind of metal surface.

Quantization is not a strange, remote idea that only matters in laboratories. It is happening in every light bulb, every sunrise, and every phone screen. The world you live in is built from energy packets. The quantum revolution was a revolution in understanding the actual structure of the energy that fills every ordinary day.

You now have a working mental model of what a quantum is: a fixed, indivisible chunk. The smallest possible amount of energy that can be emitted or absorbed at a given frequency. Atoms have specific energy levels, specific steps on a staircase, and the photons they emit correspond to the steps they jump. Light comes in particles called photons, and each photon's energy is fixed by its frequency. This is why light has color, why LEDs work, why solar panels generate electricity, and why stars have different colors at different temperatures.

Diagram 2.11 - Planck and Einstein on a Timeline

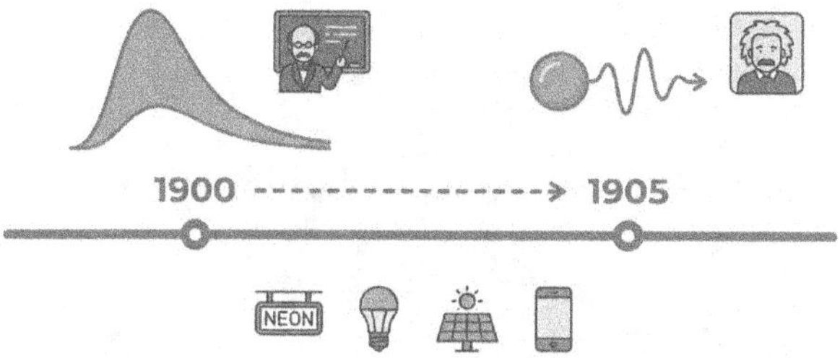

That is a genuine understanding of something real. Not a metaphor. A model of how the universe actually works at the scale of atoms and photons. It will be the foundation for everything that follows in this book, because quantum physics builds on quantization the way a house builds on its foundation. Every strange and surprising idea in the chapters ahead, from wave-particle duality to Heisenberg's uncertainty to superposition and entanglement, rests on this one central discovery: energy is not a river.

It is a rain, arriving one drop at a time, each drop complete, each drop counted.

Diagram 2.12 - The Sodium-Vapor Streetlamp

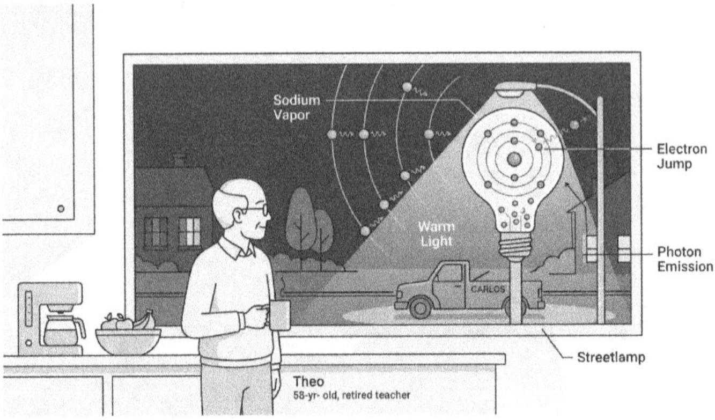

3 The Double Agent: How Light Became Both a Wave and a Particle at Once

Opening Scenario

Priya boards the 8:14 train on a Tuesday morning, backpack over one shoulder and a coffee cup in her free hand. She slides into a window seat and pops open her laptop. As the train pulls forward and picks up speed, something catches her eye. Sunlight streams through the window in long slanted beams and lands in bright, sharp bars across her laptop screen. Between each bright bar, a cool shadow. Bright. Shadow. Bright. Shadow. She tilts her head. The bars have crisp edges. Almost too perfect.

She looks down at her OLED phone on the seat beside her. She unlocks it, and the icons snap to life in vivid blues and whites. She has a graphic design assignment due this week about color and light. How does the screen make those colors? How does the sunlight separate itself into such clean bars? She doesn't think of either question as a physics question. She thinks of them as design. But both questions point to one of the strangest and most thoroughly tested discoveries in the history of science.

Light, it turns out, is a double agent. For two hundred years, scientists argued bitterly about whether light was a wave or a tiny stream of particles. The argument was settled twice, once in each direction, and then had to be reopened because the real answer is stranger than either side imagined. Light is both. Not sort of both. Genuinely, completely, provably, both at the same time. And the experiment that proved it is one you can hold in your mind in less than an hour.

Diagram 3.1 - Priya on the Train

Why It Matters

The double-slit experiment sits at the center of this chapter, and physicists still regard it as the deepest mystery in all of science. Richard Feynman, one of the most celebrated physicists of the twentieth century, called it the central mystery of quantum mechanics. He said it contains the only real quantum weirdness, and that everything else in quantum theory is just a variation on the same puzzle. That is not a small claim.

Here is what makes the experiment so unsettling. When you shoot a single particle, one electron or one photon of light, through a barrier with two tiny slits in it, and you don't try to figure out which slit it went through, the particle somehow passes through both slits at once and interferes with itself on the other side. The resulting pattern on the far wall is a wave pattern, not a particle pattern. And then, in the strangest twist of all, if you set up a detector to watch which slit the particle goes through, the wave pattern disappears. The act of observation changed the result.

This is not a theoretical quirk confined to a laboratory. Every pixel on Priya's OLED screen is produced by quantum effects that trace back directly to this experiment. Every transistor in every computer chip depends on understanding that electrons have wave properties. Electron microscopy images individual atoms by using the wave nature of electrons. You interact with the consequences of the double-slit experiment every single day. And at a deeper level, this experiment tells us the universe does not decide what a particle is doing until it is forced to. That is not a figure of speech. It is a literal description of what the evidence shows.

This chapter walks you through the experiment step by step, the way a detective lays out clues. We start with waves, then with particles, then with light, then with the single-particle surprise. We look at what happens when you try to peek at the answer. And we end with what all of this means for the way you see the world.

Diagram 3.2 - Why the Double-Slit Matters

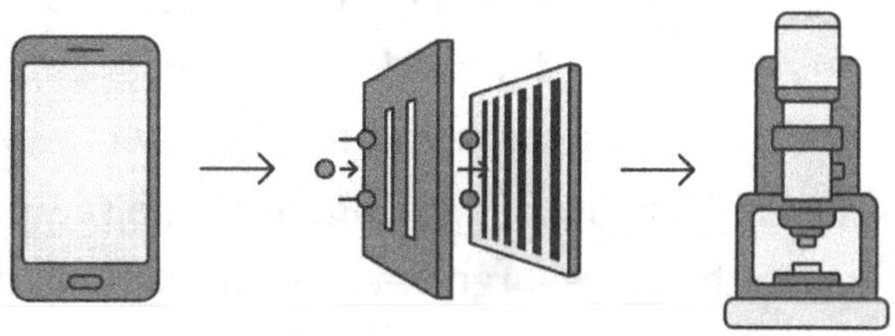

The Old Argument: Wave or Particle?

The argument about the nature of light is very old. Isaac Newton, who worked in the late 1600s and early 1700s, believed light was made of tiny particles. He called them corpuscles, which means "small bodies" or "little things". His reasoning made sense at the time. Light travels in straight lines. It predictably bounces off mirrors, like a ball bouncing off a wall. If you block part of a light beam, you get a sharp shadow. These all look like particle behaviors. Waves don't travel in straight lines the way light does, Newton argued, so light must be particles.

On the other side of the argument stood Christiaan Huygens, a Dutch scientist who was a contemporary of Newton. Huygens argued that light was a wave, like ripples spreading across a pond. His model also made sense. It explained how light bends slightly when it passes through glass, and it had a certain mathematical elegance. But Newton was Newton. He was the most influential scientist alive. His particle model dominated for about a hundred years, mostly on the strength of his reputation.

Then, in 1801, a scientist named Thomas Young did an experiment that seemed to settle the argument decisively in favor of waves. His experiment involved two very narrow slits and a beam of light, and what he saw on the far wall made the wave interpretation impossible to ignore. We will get to that experiment in detail shortly. The point for now is that after Young's result, the scientific community accepted that light was a wave. Not a stream of tiny pellets. A wave.

That conclusion stood for about a century. Then, in 1905, Einstein published a paper explaining the photoelectric effect. When light hits certain metals, it knocks electrons out of them. But low-frequency light could shine all day without knocking out a single electron, while high-frequency light knocked them out immediately, even in tiny amounts. This made no sense if light was a continuous wave. Einstein's explanation was radical. Light comes in packets, he said—small, indivisible packets of energy. One packet hits one electron, and either has enough energy to free it or doesn't. The frequency determines the packet's energy. This was the birth of the photon. Newton's particle idea was brought back from the dead.

So by the early twentieth century, science had two mountains of evidence pointing in opposite directions. Wave experiments showed that light behaves like a wave. Particle experiments showed that light behaves like a stream of photons. Both bodies of evidence were solid. Neither was going away. The scientific community was stuck with a contradiction, and it would take quantum mechanics to explain why both sides were right.

Diagram 3.3 - The Two-Sided Debate

The Double-Slit Setup in Plain English

Before we run the experiment, let's set the stage clearly to avoid confusion. The double-slit experiment has three parts: a source, a barrier, and a screen. The source fires something toward the barrier. The barrier has two narrow slits cut into it, side by side. Whatever passes through the slits hits the screen at the back. We observe what pattern forms on the screen.

That's really all there is to it, physically. A source, a barrier with two slits, and a screen to catch whatever arrives. The source can be many things: a beam of light, a stream of electrons, or even individual particles fired one at a time. The screen records where things land. The pattern that forms on the screen, over many shots or during a long exposure, tells us what we're dealing with. Is it behaving like a wave, or like a particle? The screen will tell us.

Now, before we send light through, let's run two much simpler thought experiments to build up the right intuition. The first uses water waves. The second uses bullets. If we understand what waves do and what particles do when they hit a double-slit barrier, we will immediately see why the result with light is so surprising.

Imagine you are standing at the edge of a calm pond. You drop a stone. Ripples spread out in expanding circles. Now imagine a wall stretching across the pond with two narrow gaps in it. The ripples reach the wall. Most are blocked. But some ripple energy squeezes through each gap. On the other side, two new sets of circular ripples spread out and run into each other. Where two crests meet, the water rises high. Where a crest meets a trough, they cancel, and the water is almost flat. This mixing is called interference. The result on the far bank is not two simple humps. It is an alternating pattern of high ripples and flat areas that marches across the bank.

That pattern, bright bands where waves add up and dark gaps where they cancel, is the fingerprint of a wave. If you see that pattern, you are dealing with a wave. If you don't see it, you're probably dealing with something else.

Diagram 3.4 - Pond Ripples Through Two Gaps

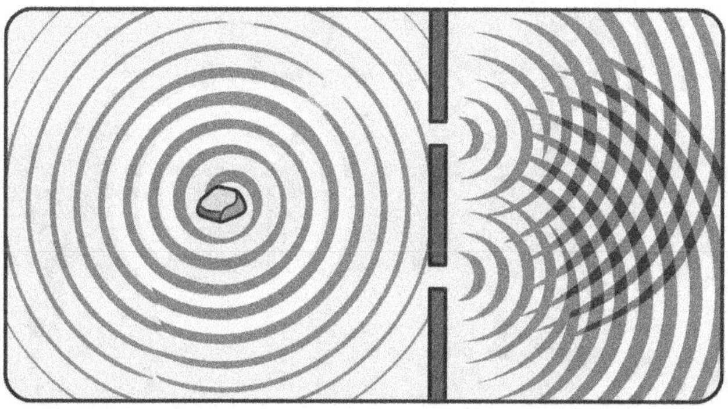

Now switch to a completely different scenario. Imagine a machine gun firing bullets at a solid wall with two holes in it. The bullets are solid objects. They travel in straight lines. Some hit the wall and stop. Others pass through one hole or the other. The ones that pass through hole one spray out in a fan pattern centered on hole one. The ones that pass through hole two spray out in a fan pattern centered on hole two. On the back wall, those two fan patterns overlap, and you end up with, roughly, two stripes of bullet holes. A hump of hits near where hole one points, a hump of hits near where hole two points, and relatively few hits in between. No ripple pattern. No alternating bright and dark bands. Just two simple clumps. This is the fingerprint of particles.

So the rule is simple. Waves give you a many-banded ripple pattern. Particles give you two simple stripes. When Thomas Young aimed light at a double slit in 1801, he was essentially asking: which one does light give me? He was expecting two stripes. What he got was very, very different.

Diagram 3.5 - Bullets Through Two Holes

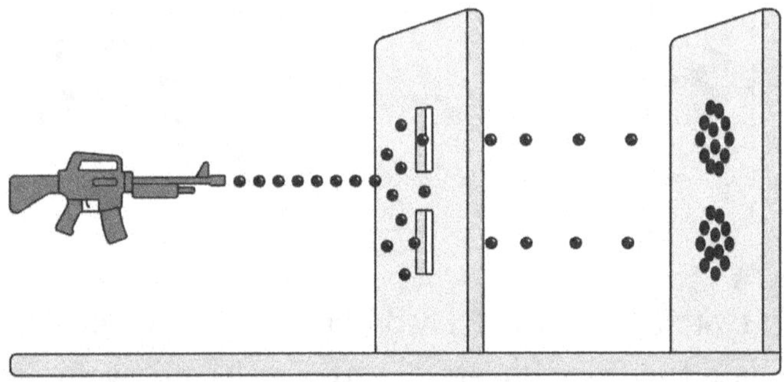

What Happens with Waves

Let's slow down and really understand wave interference, because it is the heart of everything that follows. When two waves meet, something very specific happens based on where each wave is in its cycle at the moment of meeting. A wave has peaks and troughs. A peak is the high point of a wave. A trough is the low point. If two peaks arrive at the same point at the same moment, they add together. The result is a bigger peak. This is called constructive interference. The waves reinforce each other.

The opposite can also happen. If a peak from one wave arrives at the same point where a trough from another wave is arriving at the same moment, they cancel each other out. The result is nothing. Flatness. Silence. This is called destructive interference. The waves erase each other.

In between those two extremes, partial adding and partial canceling produce waves of intermediate size. The overall result when two wave sources are side by side, like our two pond ripples coming through two slits, is a standing pattern of bright bands and dark gaps. The bright bands are where the waves from both slits arrive in step, adding up. The dark gaps are where the waves arrive exactly out of phase and cancel each other out.

The spacing of the bright bands and dark gaps depends on the wavelength of the wave, which is just the distance between successive peaks, and on the distance between the two slits. Longer wavelengths make wider bands. Shorter wavelengths make narrower bands. This is not a vague prediction. It is precise and testable. You can calculate exactly where each bright band will fall before you even run the experiment, and then go check. Waves always produce this pattern. Always.

Think about noise-canceling headphones. The headphones have a microphone that picks up the sound around you. A circuit generates a new sound wave that is an exact mirror image of the incoming noise, peak for trough and trough for peak. The two sounds cancel via destructive interference. The headphones use wave interference on purpose, as a tool, to give you silence. It works reliably every time because wave interference is not a quirk. It is a fundamental property of anything that behaves like a wave.

So when Thomas Young sent light through two slits and looked at the screen on the far side, the pattern he saw was unmistakably the pattern of wave interference. Alternating bright fringes and dark fringes, spaced exactly as the wave theory would predict, using the known wavelength of light. The experiment was clean, repeatable, and absolutely decisive. Light was behaving like a wave. There was no two-stripe result. There was no particle footprint. There was a ripple pattern, marching across the screen in perfect stripes.

Diagram 3.6 - Wave Interference Through Double Slit

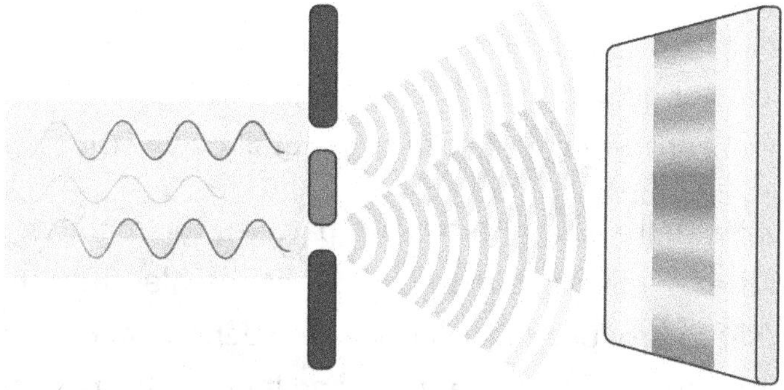

What Happens with Bullets

Let's hold that wave result in mind and contrast it sharply with what happens when genuine particles go through a double slit. This comparison is essential because the surprise of the experiment with light, and later electrons, only registers fully when you have seen what ordinary particles do.

Imagine your particle gun is firing marbles, or sand grains, or any small solid object. The objects travel in straight lines. When they reach the barrier, each one either hits the wall and stops, or it passes through one of the two slits. Not both. One or the other. It's a solid object. It can only be in one place at a time. It goes through the left slit or the right slit.

After passing through a slit, the marble continues moving in roughly the same direction. It spreads out slightly because the edges of the slit deflect it, but it mostly continues forward. It lands on the screen somewhere in a region corresponding to the slit it came through. Add up thousands of marble shots, and you see two clusters building up on the screen. One cluster behind the left slit. One cluster behind the right slit. Maybe with a little smearing at the edges, but fundamentally: two stripes.

Crucially, there is no interference pattern. The marbles from the left slit don't interact with the marbles from the right slit to produce cancellations between them. Each marble is its own separate thing, and it doesn't care about the other slit. It went through its own slit, it landed where it landed, and that's the end of it. Particles are local. Particles are definite. Each one has a single, definite position at all times.

This is the baseline against which the experiments with light and with electrons will be measured. If we are dealing with particles, we should see two stripes—no more, no less. Two stripes are the correct answer if light is particles. An interference pattern with many bands is the correct answer if light is a wave. Thomas Young ran the experiment, and the answer was not two stripes. It was a many-banded interference pattern. At the time, this seemed to decisively prove that light was a wave. Newton was wrong. Huygens was right.

But then Einstein came along a hundred years later with the photoelectric effect and the photon, and suddenly the particle picture had solid experimental support too. How could both be right? The two results seemed to contradict each other flatly. An interference pattern with many bands could only be produced by a wave. But the photoelectric effect could only be explained by particles. Both experiments were rigorously done. Neither could be dismissed. The contradiction was real.

Diagram 3.7 - Two-Slit Comparison: Waves vs Bullets

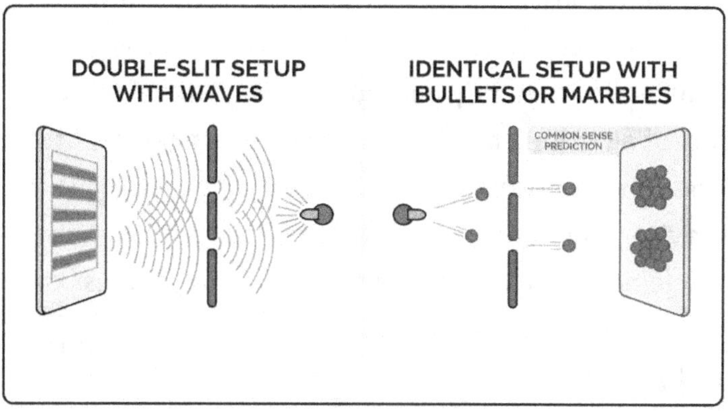

What Happens with Light

Thomas Young's experiment in 1801 was elegant in its simplicity. He took a beam of sunlight, sent it through a tiny pinhole to create a concentrated source, and then directed the light toward a card with two very thin slits cut side by side. On a screen some distance behind the card, he observed what formed.

What he saw were stripes. Not two stripes, but many stripes. Bright and dark stripes alternating evenly across the screen, with the brightest stripe in the center and the stripes gradually dimming toward the edges. The spacing of the stripes matched almost exactly what you would predict if light had a particular wavelength and was behaving like a wave spreading from two sources. The math worked out precisely. The wave picture fit perfectly.

Young tried it with different colors of light and found that different colors produced stripes of varying spacing. Red light, which has a longer wavelength, produced wider-spaced stripes. Blue light, which has a shorter wavelength, produced more tightly packed stripes. White light, which is a mix of all colors, produced a rainbow-smeared pattern of stripes. Every observation was consistent with wave behavior, and none with particle behavior. Two-stripe particle behavior would give two stripes regardless of color. What Young saw was unmistakably wave interference.

This result was not widely accepted at first, partly because of Newton's enormous reputation. But over the following decades, more experiments confirmed Young's result, and eventually even the most stubborn holdouts had to admit that light was behaving like a wave. By the middle of the nineteenth century, the wave theory of light was essentially settled science.

The same stripes Priya sees in her window are a version of this. Sunlight passing through the slats of window shades or through narrow gaps between objects creates interference-like effects when the geometry is right. The perfectly spaced bars of light on her laptop screen come from a different, simpler mechanism. Still, the principle is the same: light spreading and overlapping and producing structured patterns of brightness and shadow. This is not special. This is just what light does, because light is a wave.

And yet. Einstein's 1905 paper on photons would upend everything again. Light is a wave. Light also comes in packets. Both are true. How can that be? That is the question the next two sections answer, and the answer requires us to do something that will initially feel impossible: fire light, one photon at a time, through the double slit, and watch what happens.

Diagram 3.8 - Sunbeam Through Window Slats

The Big Surprise: Single Particles Make a Wave Pattern Too

Now we arrive at the heart of the mystery. The next step is the one that made Feynman say what he said.

We established that light going through a double slit produces an interference pattern. You could still tell yourself a comfortable story at this point. Maybe the photons in the beam are interacting with each other. Maybe many overlapping photons create interference. None of those stories is correct, but they have a certain appeal. They keep particles being particles and waves being something separate.

Those comfortable stories collapse completely when you experiment with one particle at a time.

Here is the experiment. You dial down the light source so finely that you send one photon at a time through the apparatus. You send a photon and wait. Then another. You record where each one lands. After one photon, you see a single dot. After ten, you see ten dots scattered around. After a hundred, a vague cluster. But after thousands of photons, each sent alone with no other photon present during its journey, a pattern emerges. The dots are not forming two stripes. They are forming the familiar multi-banded interference pattern. Bright zones where many dots cluster. Dark zones where almost no land—the wave fingerprint, building up dot by dot.

Let that sit for a moment. Each photon was alone. Nothing to interfere with. And yet it landed in exactly the places a wave would land, and avoided the places a wave would cancel. How did a single particle produce a pattern that only makes sense if a wave went through both slits simultaneously?

The same experiment has been done with electrons, one at a time. Each electron has a definite mass and a definite electric charge. It is as particle-like a thing as anything in physics. And yet, when you send thousands of electrons through one at a time, the same interference pattern builds up. Dot by dot. Each electron arrives alone. And still a wave pattern.

The experiment has also been performed with atoms, molecules, and buckyballs, which are soccer-ball-shaped molecules composed of 60 carbon atoms. Each one is sent through individually, and it still builds up an interference pattern. Wave behavior is not limited to photons and electrons. It is a feature of matter itself at the small scale.

Carlos, back at his auto repair shop, would find this completely baffling. He fixes things by tracking them. A car part is in one place, then another. You can follow it. But an electron going through the double slit has no definite path. It doesn't go through the left slit or the right slit. The interference pattern can only be produced if the wave-like aspect of the particle explores both paths simultaneously. The electron is not a tiny marble with a trajectory. It is something that has no definite location until it lands on the screen.

Diagram 3.9 - Single-Electron Interference Pattern Building Up

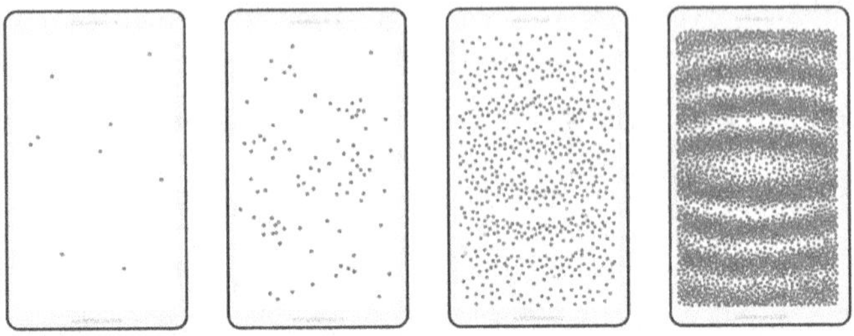

The physicist's way of saying this is that before detection, the particle exists in a superposition of all possible paths. Its wave function spreads out through space and passes through both slits. On the other side, it interferes with itself, creating regions of high probability and near-zero probability. When the particle hits the screen, it snaps to a definite location in one of the high-probability regions. Repeat this enough times, and the statistical buildup traces out the interference pattern.

Notice what this means for the old argument. Newton was not entirely wrong. A photon arrives at the screen as a single point, with definite energy and momentum. But Huygens was not entirely wrong either: the process of getting there is governed by wave interference. The particle and the wave are two aspects of the same entity, one governing how it travels and the other governing where it arrives.

Diagram 3.10 - Wave-Particle Duality Illustrated

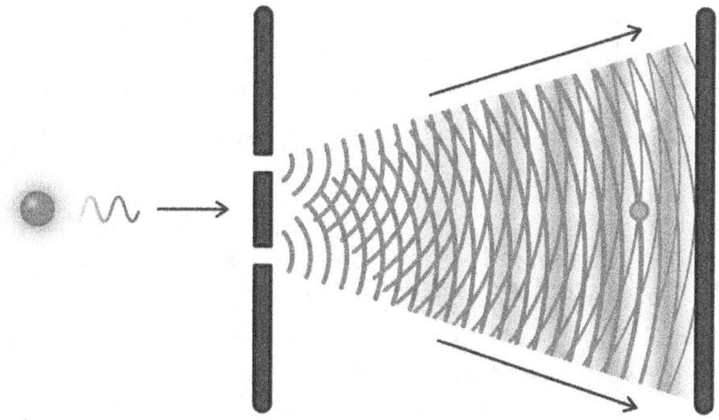

The Stranger Twist: Watching Changes the Result

If the single-particle interference result made you uncomfortable, the next step will be genuinely unsettling because there is a way to make the interference pattern disappear. And all it takes is knowing.

Physicists naturally wanted to understand how a single particle manages to explore both slits. The obvious move was to find out which slit each particle actually went through. They set up a detector right at the slits, a tiny device that registers whether a particle passes through the left or right slit without stopping it. The idea was to keep the experiment running as before, but now also collect which-path information.

The result was stark. The moment you turn on the which-path detector, the interference pattern vanishes. The screen shows two stripes, one behind each slit, exactly like the marble experiment. The wave behavior is gone. The particles now behave as if they really did pass through one slit or the other, like ordinary little objects.

Turn the detector off. The interference pattern comes back.

Turn it on again. Two stripes.

Off. Interference pattern.

The detector doesn't have to detect every particle. If you turn it on but don't look at the results, the pattern still disappears. The issue is not whether a human reads the data. The issue is whether the which-path information exists anywhere in the universe, even in principle. If it is physically available, even unread, the interference pattern is gone. If it is not available, the pattern comes back.

This is the result that stops people cold. The particle responds not to a conscious observer but to the presence of path information in the physical world. Before measurement, the particle's wave function stretches through both slits and interferes with itself. The moment the which-path measurement is made, the wave function collapses to a single path, and interference disappears.

There have been clever variations of this experiment to probe how deeply it runs. In one version, scientists found a way to erase the which-path information after the particle had already passed through the slits. When the path information was erased, the interference pattern came back, even though the particle had already traveled most of the way. This is called the quantum eraser experiment. Physicists continue to argue about exactly how to interpret it.

What physicists do NOT say is that the particle knows a person is watching. The interference pattern is not affected by consciousness. It is affected by physical interaction. A detector is a physical device. When a particle interacts with it, information is transferred, and that transfer changes the outcome. The word observer in quantum mechanics means any physical system that registers information about the particle. It does not mean a mind or a person. It means a physical interaction that leaves a record.

Diagram 3.11 - Detector On vs Detector Off

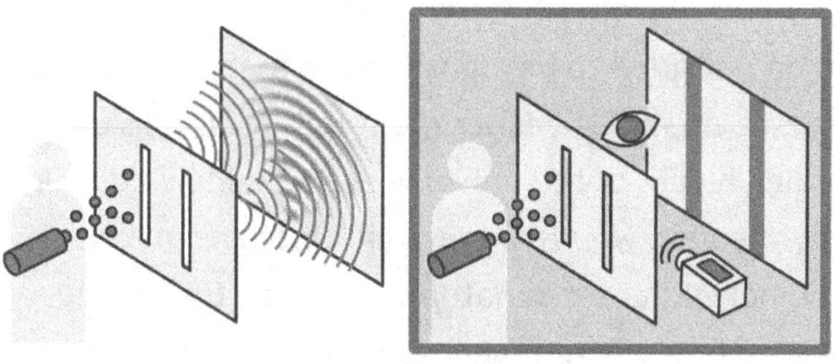

There is a technical reason why measurement destroys the interference. To detect which slit a particle went through, the detector must interact with the particle, even very gently. That interaction entangles the particle with the detector. Once entangled, the two possible paths carry different information about the detector state and can no longer be cleanly combined. The interference is washed out. This process is called decoherence.

Decoherence is also why everyday objects don't behave quantum-mechanically. Cars and coffee cups are constantly interacting with their environment through photons, air molecules, and thermal vibrations. Each interaction records information about the object. Its quantum coherence is destroyed almost instantly. Only very small, very isolated systems maintain wave-like behavior long enough for effects like the double-slit pattern. Quantum weirdness lives in the gaps where nature hasn't been forced to decide. Once a system is large enough and connected enough, those gaps close.

What This Means For You

You do not need to be a physicist to feel the significance of what the double-slit experiment tells us. It tells us that the building blocks of the universe are genuinely different from anything we see and touch. Not just smaller. Genuinely different in nature. A particle that produces a wave pattern while traveling alone through space is not a tiny marble. It is something we have no everyday word for.

But the everyday consequences are everywhere, once you know where to look.

Priya's OLED phone screen works through layers of organic molecules engineered to allow electrons to tunnel through them and emit light at specific frequencies. That engineering depends on understanding the wave nature of electrons, which is exactly what the double-slit experiment revealed. Without it, there would be no transistors, no chips, no phones, no internet.

The electron microscope is another direct consequence. The wavelength of visible light limits the performance of light microscopes. Electrons have a much shorter wavelength, allowing electron microscopes to image individual atoms. This technology is used in research labs, materials science, and chip manufacturing, and it relies entirely on electrons behaving as waves.

The colors on a soap bubble or an oil slick are produced by wave interference, too. Light reflects off both surfaces of a thin film. The two reflected beams interfere. Some wavelengths add up and bounce back strongly, others cancel out, producing the shimmering rainbow you see on a soap film in sunlight. Every time you see that, you are watching Thomas Young's experiment happen in real life.

And the deepest implication: the double-slit experiment tells you that reality at the smallest scale is wave-like until it is measured. The interference pattern proves that both paths were genuinely physical before detection. This is not a limitation of our instruments. It is a feature of the universe.

Theo, the retired history teacher, would find a parallel here. In history, there were always paths not taken. In the quantum world, those other paths are physically real right up until the moment of measurement. Then they close. The universe, at the smallest scale, keeps multiple possibilities open at once. Measurement is what forces a choice.

Diagram 3.12 - Everyday Connections to Wave-Particle Duality

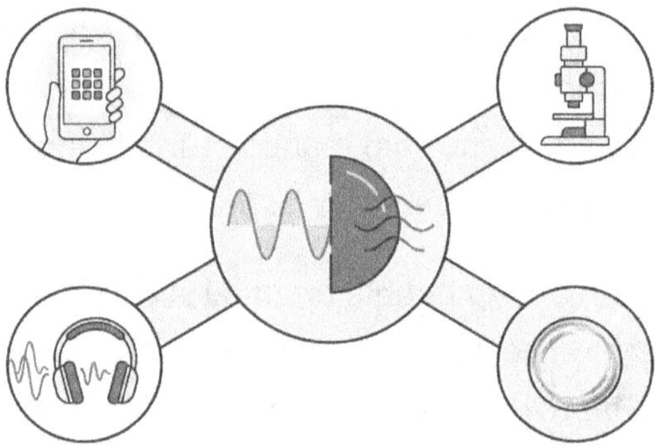

Takeaway

The double-slit experiment is not a party trick or a curiosity. It is a window into how the universe actually works at the level of its smallest pieces. What it shows is this: the things that make up matter and light are neither waves nor particles in the everyday sense of those words. They are genuinely new, something for which human language has only imperfect tools. They travel as waves, spreading and interfering with one another. They arrive as particles, landing at single points. And crucially, the act of measuring which path they took is not a passive act of observation. It is an interaction that changes the system's physical state and erases its wave behavior.

Priya, by the time her train pulls into the campus station, might look at the bars of sunlight across her laptop and think about Young's experiment. She might unlock her OLED phone and think about the wave nature of the electrons that are producing each pixel. She might notice the slight shimmer on the window glass and see wave interference instead of a smudge. None of these requires her to solve an equation or memorize a formula. They require her to hold one idea clearly: light is a double agent, and so is everything else at the quantum scale. Wave on the way in. Particle on the way out. The universe keeps its options open for as long as it can.

That is the core idea of wave-particle duality. It is also the first step toward understanding the deeper strangeness of quantum mechanics, where things can be in two states at once, and the act of asking a question is inseparable from the answer you get. We will get to both of those in the chapters ahead.

What you have now is something better than a simple answer to a simple question. You have a genuine grasp of why the question is hard, why it took two hundred years and some of the most careful experiments ever performed to arrive at an answer, and why the answer is genuinely startling even to scientists who have spent careers working with it. That understanding, that earned sense of wonder, is exactly what Feynman was pointing at when he called this the central mystery of quantum mechanics. Now it is yours too.

You can describe the double-slit experiment from memory. You know what the interference pattern means and why it is a fingerprint of wave behavior. You know that single particles produce that pattern even when fired alone. You know that watching which slit they go through erases the pattern, not because of magic, but because measurement is a physical interaction that changes the system. You know that this is not a misunderstanding or a measurement error. It is the literal behavior of matter and light at the quantum scale. Hold that. The next chapter will take it further.

4 The Fog of Certainty: What Heisenberg Actually Discovered

Opening Scenario

Maya is already running late. She grabs her phone off the kitchen counter, opens her navigation app, and watches the blue dot blink to life somewhere in the middle of her street. It is not a perfect dot. It is a small circle, a soft blue halo that pulses gently around her approximate position. The circle tells her something honest: the app knows roughly where she is, but not exactly. A few meters of fuzziness live inside that ring. She does not think much of it. She taps the fastest route and heads for the door.

Halfway to the subway, she notices the speed readout too. Her phone says she is walking at about 4.8 kilometers per hour. The number updates every second, and it nudges up and down by small amounts even when her stride feels perfectly steady. There is also fuzziness in the speed estimate. Location is a bit uncertain. Speed is a bit uncertain. Maya mentally files this under "technology is imperfect" and keeps walking.

But here is what Maya does not know yet. That fuzziness is not a flaw. It is not a sign that her GPS chip needs a firmware update or that her phone's antenna is too small. There is a deeper reason why position and speed resist being nailed down at the same moment, and it has nothing to do with engineering. It has to do with the way the universe is built. Today, Maya is going to learn that Werner Heisenberg discovered this same fuzziness sitting at the heart of everything, long before smartphones existed, and that what he found changed physics forever.

Diagram 4.1 - Maya and the GPS Uncertainty Circle

USING UNCERTAINTY INLOCATION ESTIMATION AND SPEED

Why It Matters

The uncertainty principle is probably the most famous result in all of quantum mechanics. It is also one of the most misunderstood. Most people who have heard of it picture a scientist poking at a tiny particle with a blundering instrument, accidentally knocking it off course. The idea seems to be that if we had better tools, more delicate probes, faster computers, we could eventually measure everything precisely. The fuzziness in that reading is our problem. It is a limitation of our cleverness, not of nature.

Heisenberg's actual discovery says something far stranger. The fuzziness is not in our instruments. It is in the universe itself. There is no instrument, however perfect, that could ever simultaneously pin down both the exact position and the exact momentum of a particle. Not because the instrument disturbs the particle. Because position and momentum are not two independent facts about a particle. They are paired, linked, and traded against each other, like the opposite ends of a seesaw. The more precisely one is defined, the less precisely the other can be, always, for every particle, in every corner of the universe.

This matters to you even if you never think about subatomic particles. It explains why atoms are stable. It explains why electrons do not spiral into the nucleus and collapse everything into a tiny point of nothing. It explains why lasers work, why the transistors in your phone behave the way they do, and why even space is never truly empty. The uncertainty principle is not an annoying footnote at the bottom of physics. It is one of the load-bearing walls.

After this chapter, you will be able to look at Maya's GPS circle and see something different. You will not see a technology problem. You will see a small everyday echo of a rule that runs all the way down to the foundations of existence.

The Old Promise: Newton Said We Could Know Everything

To understand why Heisenberg's discovery was such a shock, you need to feel how confident physics was before it arrived. For two and a half centuries, the dominant picture of reality was Isaac Newton's. His laws of motion were precise, elegant, and breathtakingly successful. They explained the orbits of planets, the arc of a cannonball, the swing of a pendulum, and a thousand other phenomena with startling accuracy.

Newton's world was a world of certainties. If you knew the exact position of every particle in the universe at a single moment, and you knew the exact speed and direction each particle was moving, then in principle you could calculate exactly where everything would be at any future moment. The universe was a giant clockwork machine. It had no secrets. It had no fog. Every measurement had a definite answer, and if you got a fuzzy answer, that was purely because your ruler was not fine enough, not because nature itself was vague.

A French mathematician named Pierre-Simon Laplace pushed this idea to its logical limit. He imagined a being of perfect intelligence, knowing the position and velocity of every particle in existence. For such a being, Laplace wrote, nothing would be uncertain, and the future, just like the past, would be present to its eyes. This imaginary being came to be called Laplace's Demon. It was the ultimate expression of Newtonian faith: the universe is completely knowable, if only we are clever and careful enough.

This vision was beautiful. It was also, as physicists would discover in the early twentieth century, completely wrong, at least at the level of the very small.

Diagram 4.2 - Newton's Sharp Pencil Prediction

OVERTURNING CLASSICAL PREDICTABILITY: BEYOND CLOCKWORK ORBITS

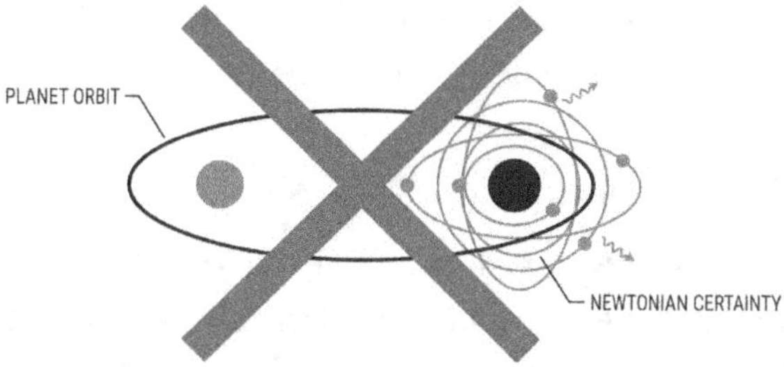

NEWTONIAN DETERMINISM CANCELLED: QUANTUM MECHANICS INTRODUCES FUNDAMENTAL UNCERTAINTY.

What made Newtonian physics feel so reliable was its perfect track record at human scales. Drop a ball off a building, and Newton's laws tell you exactly when it will hit the ground. Launch a rocket, and those same laws guide it to the moon. At everyday scales, the clockwork picture works extraordinarily well.

The trouble only becomes visible when you zoom in far enough, down past molecules, past atoms, past the nucleus, into the strange territory where particles exist as smears of possibility rather than hard little balls in defined locations. At that scale, Newton's sharp predictions dissolve into something softer.

Werner Heisenberg was a young German physicist in the 1920s, working on the puzzling behavior of electrons in atoms. The uncertainty principle, which he published in 1927, was not a guess. It was a rigorous result that followed inescapably from the new framework physicists were building to describe the quantum world. In plain terms, there are pairs of physical properties such that the more precisely you know one, the less precisely you can know the other. Not because of instrument limitations, but because of the deep structure of reality.

Heisenberg's Surprise: A Built-In Fog

The most famous pairing is position and momentum. Momentum is essentially a particle's mass multiplied by its velocity. For everyday purposes, you can think of it as how fast the particle is moving and in what direction. Heisenberg showed that position and momentum form a complementary pair. You cannot know both to arbitrary precision simultaneously.

Think of it this way. Imagine you want to locate a particle very precisely. You want to know exactly where it is. To do that, you essentially need to pin it down in space, confine it to a very small region. But the moment you confine a particle to a small region, something strange happens to its momentum. It becomes wildly uncertain. The particle's speed and direction can be almost anything. The range of possible momenta is enormous.

Now go the other way. Imagine you want to know a particle's momentum very precisely. You want to know its speed and direction with great accuracy. The trade-off runs in reverse. Now the particle cannot be localized to a small region. Its position becomes spread out, smeared across a large area. It is sort of everywhere at once, because that is the cost of having a well-defined momentum.

This is the fog that Heisenberg discovered. It is not grim on your glasses. You cannot wipe it away. It lives within the mathematics of quantum mechanics, and no improvement in measurement technology can remove it, because it is not about measurement technology at all.

Diagram 4.3 - The Position-Momentum Trade-Off

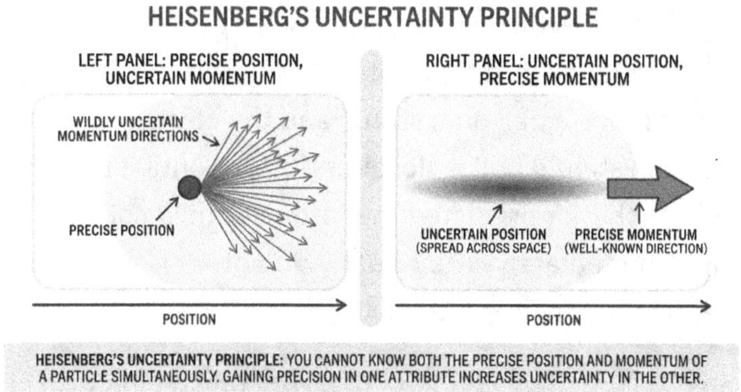

There is a helpful way to think about this. In quantum mechanics, particles are described not as little hard spheres but as waves. A wave has an important property: the more tightly you squeeze a wave into a small region of space, the more frequencies you need to build it. A pure single frequency spreads out over all of space. You cannot have a wave that is both confined to a tiny region and made of a single pure frequency at the same time.

In quantum mechanics, the frequency of a particle's wave is connected to its momentum. So when you say a particle has a very precise momentum, its wave spreads out everywhere, and the particle's position is completely undefined. When you say the particle is tightly localized in space, its wave requires mixing many different frequencies, meaning many different momenta, and the momentum is wildly uncertain. The uncertainty principle is a consequence of what waves do. And particles, as we saw in Chapter Three, are waves.

This also helps explain one of the most puzzling questions about atoms. Why do electrons not simply spiral into the nucleus? Opposite charges attract. An electron orbiting a nucleus is like a ball on a string, and the electromagnetic attraction should pull it closer and closer until it crashes in. Classical physics predicted that atoms should collapse almost immediately. They clearly do not. Heisenberg's uncertainty principle is a major reason.

If an electron were confined to a very tiny region near the nucleus, its position would be very well defined. But a precisely defined position means wildly uncertain momentum. That means enormous kinetic energy, energy of motion. The electron essentially kicks back hard the moment you try to pin it down too tightly. It is as though the act of confinement itself energizes the electron and pushes it outward. The two effects, the electromagnetic attraction pulling inward and the quantum uncertainty kicking outward, reach a natural balance. That balance is the stable atom. Heisenberg's fog is what holds ordinary matter together.

Diagram 4.4 - Why Electrons Don't Spiral In

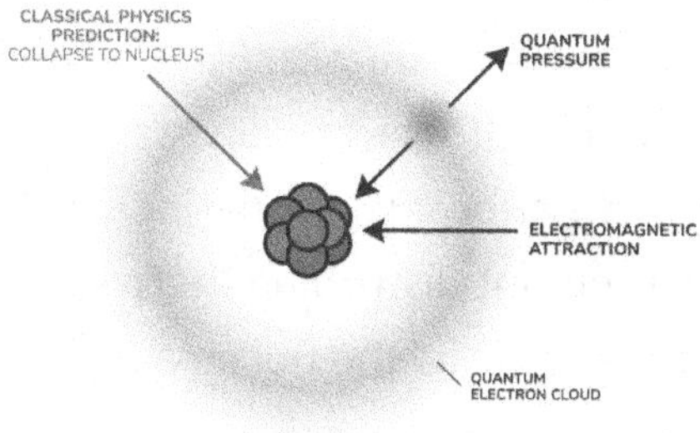

WHY ELECTRONS DON'T SPIRAL IN
Inward electromagnetic attraction is balanced by outward quantum pressure, preventing the electron cloud from collapsing into the nucleus.

Why It's Not Just Clumsy Measurement

This is the point that trips up almost everyone. The uncertainty principle says that when we measure a particle's position, we disturb it, and that disturbance scrambles its momentum. If only we could measure more gently, we could know both. This interpretation matches our everyday experience. Poking something moves it.

Heisenberg himself, early in his thinking, reached for this kind of explanation. He imagined a hypothetical super-microscope that used gamma rays to spot an electron's exact position. Gamma rays have very short wavelengths so that they can pinpoint a location precisely. But they also carry a lot of energy. When a gamma ray photon hits an electron, it delivers a sharp kick that scrambles the electron's momentum. The sharper your position measurement, the more violent the momentum disturbance.

This thought experiment, known as Heisenberg's gamma-ray microscope, is often used in popular books. It makes the principle sound like a practical problem: use gentler light, and the trade-off shrinks. But Niels Bohr pointed out to Heisenberg that the microscope story is actually misleading. It describes what physicists call the observer effect: measurement disturbing the system. That is real. But it is not the same as the uncertainty principle.

The genuine uncertainty principle is deeper than any measurement disturbance. It says that an electron does not have a precise position and a precise momentum at the same time before you measure it, regardless of how you measure it. The fuzziness is not caused by your probe poking the particle. The particle never had a well-defined position or momentum to poke in the first place. The indefiniteness is intrinsic. It is baked into the particle's nature.

Diagram 4.5 - Observer Effect vs. True Quantum Uncertainty

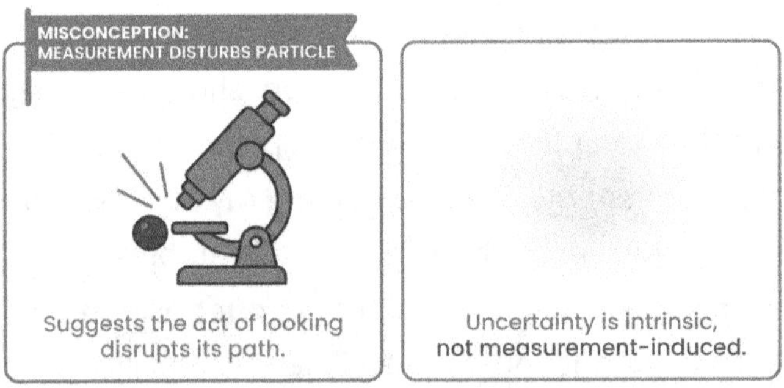

Understanding Heisenberg's Uncertainty Principle.

Think of it this way. Imagine you have a photograph of someone running. If the shutter speed is slow, the runner appears blurry. You might be tempted to say the blurriness is a defect of the camera. But in the quantum world, there is no sharp runner underneath the blur. The blur is the runner. The particle does not secretly have a definite position and momentum that your clumsy instruments fail to measure. It genuinely exists in a state that has no definite value for both at once.

This is why physicists use the term ontological to describe quantum uncertainty, rather than epistemic. Epistemic uncertainty is uncertainty in our knowledge, a gap between what is and what we know. Ontological uncertainty is uncertainty about reality itself, a state of affairs in which there is no definite fact to be known. Classical physics had only epistemic uncertainty. Quantum mechanics introduced ontological uncertainty, and the universe has never been the same since.

The distinction is not just philosophical. It has testable consequences. Physicists have conducted elaborate experiments over many decades to determine whether the uncertainty is truly ontological or whether particles secretly have definite values that our measurements cannot access. These hidden-variable tests, connected to the work of physicist John Bell in the 1960s, gave a clear answer: the uncertainty is real. Particles genuinely do not have definite simultaneous values for complementary pairs of properties. Nature is irreducibly fuzzy at its smallest scales.

Frank, if he were listening, might raise a skeptical eyebrow at this point. He would ask how something can fail to have a definite value. That instinct is completely reasonable. It comes from a lifetime of experience with ordinary objects, and for ordinary objects, it is perfectly accurate. A baseball is somewhere. It has a speed. Our intuitions are calibrated to objects composed of trillions of trillions of particles. At that scale, quantum fuzziness averages out, and the universe behaves exactly as Newton predicted. But zoom down to a single electron, and you are in a different regime, one where those intuitions do not apply.

The GPS Analogy: Sharp Location or Sharp Speed, Not Both

Let us come back to Maya and her GPS. She is standing on a city street, and the blue circle on her screen is a few meters wide. She can see roughly where she is, but the app cannot tell her she is standing exactly 14.223 meters from the north corner of the intersection. There is a range—a zone of possibility.

A conventional GPS works by measuring the time it takes for signals from satellites to reach your phone. The timing has to be extraordinarily precise because signals travel at the speed of light, and even a tiny error in timing translates into a meaningful error in position. Modern GPS can get quite accurate, often to within a few meters in everyday use and much less in specialized applications. The uncertainty circle Maya sees is partly a reflection of these real engineering limits.

But the GPS analogy carries a deeper lesson if you look at it the right way. Position and velocity are, in a sense, in tension even in the GPS. To get a good position fix, the GPS chip needs to lock onto signals and average over time. The longer it averages, the better the position estimate. But during that averaging window, you are moving. The longer you average to sharpen the position, the more your actual location has drifted. The speed estimate, meanwhile, is computed by comparing positions over time. A longer time window gives a better average speed, but loses track of where you are right now.

This is not quite the quantum uncertainty principle in action at the human scale, because quantum effects are immeasurably small for objects as large as a phone or a person. But the structure of the trade-off is the same: pinning down position and pinning down velocity pull in opposite directions. You cannot optimize both at once. The GPS is navigating a classical version of the same tension that Heisenberg found at the quantum scale.

Diagram 4.6 - Sharp Location or Sharp Speed, Not Both

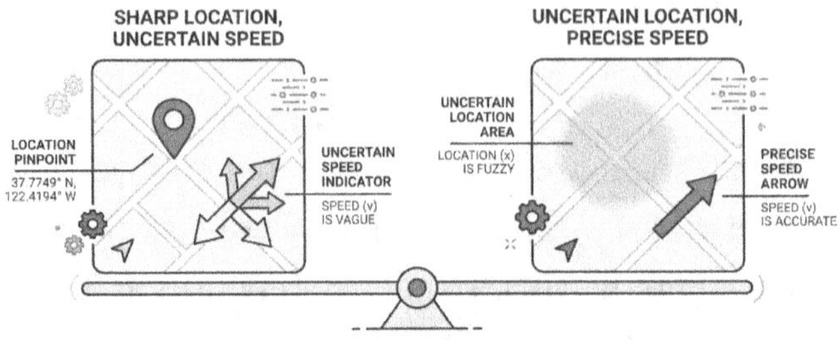

THE UNCERTAINTY PRINCIPLE IN NAVIGATION (TRADE-OFF)

Now scale that trade-off down to an electron. An electron is incomparably smaller than a phone. The quantum effects that are negligible at human scales dominate completely at electron scales. For an electron, the Heisenberg fuzziness is not a small correction or an engineering imperfection. It is the entire story. The electron genuinely does not have both a precise location and a precise momentum. Asking for both is like asking for a piece of music to be both very long and very short at once. The two properties are not just hard to measure simultaneously. They cannot both exist simultaneously.

Here is a concrete way to feel this. Suppose you want to trap an electron in a very small box. The smaller the box, the better you know the electron's position. But as the positional uncertainty shrinks, the momentum uncertainty grows. In a very tiny box, the electron races around with enormous uncertainty in momentum, meaning it can have almost any speed and direction. That furious quantum vibration is called zero-point energy. It is a direct consequence of confinement, and it never goes away, not even at absolute zero temperature.

Zero-point energy has measurable effects. At very low temperatures, liquid helium remains liquid rather than freezing because the quantum zero-point vibrations of helium atoms prevent them from locking into a crystal lattice. The uncertainty principle keeps helium liquid at temperatures colder than those at which any other substance on Earth becomes solid.

Maya's GPS circle is not just a phone feature. It is a small, everyday reminder that there are things you cannot sharpen past a certain point. At human scales, the sharpness limit is set by engineering. At quantum scales, it is set by Heisenberg's uncertainty principle.

Energy and Time Have Their Own Trade-Off

Position and momentum are the most famous complementary pair, but they are not the only ones. Energy and time form a similar partnership, and the consequences are, if anything, even stranger.

The energy-time version of the uncertainty principle says: the more precisely you want to measure the energy of a system, the longer you need to observe it. A measurement made over a very short time interval yields an energy value with a large built-in uncertainty. This fuzziness is invisible in everyday life. Measure the energy of a battery, and you can take as long as you like. The energy-time trade-off only becomes dramatic at the tiniest fractions of a second that matter in the quantum world.

Think about a short-lived excited atom. An atom can absorb energy and jump to a higher energy state. That state does not last forever. After a brief time, the atom releases the excess energy as light and returns to its ground state. The shorter the lifetime, the more uncertain the energy of the emitted light. It is the energy-time version of Heisenberg's principle at work. A light source whose atoms remain excited for only a brief moment produces light with a smeared frequency range rather than a single pure one.

Diagram 4.7 - Energy and Time: The Other Trade-Off

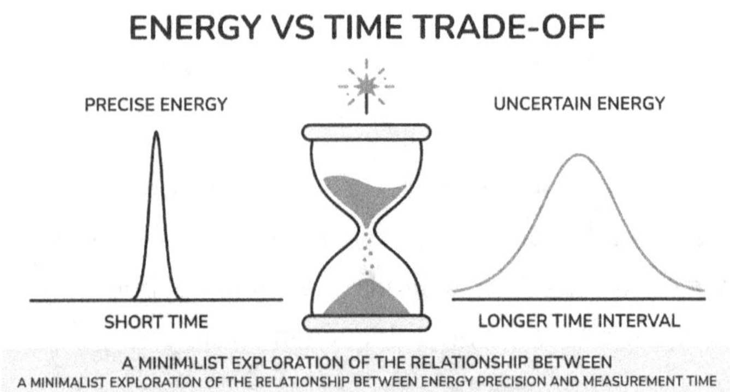

Now here is where things get truly remarkable. If you observe a region of space over an extremely short time interval, the energy uncertainty during that interval can be very large. And large energy uncertainty means that for a very brief moment, there can be enough energy to create particles, real particles, out of nothing.

These briefly appearing particles are called virtual particles. They pop into existence, exist for an incredibly short time, and then disappear again. The shorter their existence, the more energy they can borrow from the uncertainty principle's built-in fuzziness. They are not imaginary. They are not a metaphor. They have measurable effects on the real world, effects that physicists have tested and confirmed with great precision.

The most famous measurement of this effect is called the Casimir effect. Place two very clean, very flat metal plates extremely close together in a vacuum. The small gap restricts which virtual particles can form between them, while outside the plates, a fuller range can pop in and out. This creates a tiny but measurable pressure that pushes the plates together. The Casimir effect was predicted in 1948 and measured in 1997, matching theoretical predictions with remarkable accuracy. The quantum vacuum is not empty. It is alive with transient, flickering energy.

Diagram 4.8 - Virtual Particles in the Quantum Vacuum

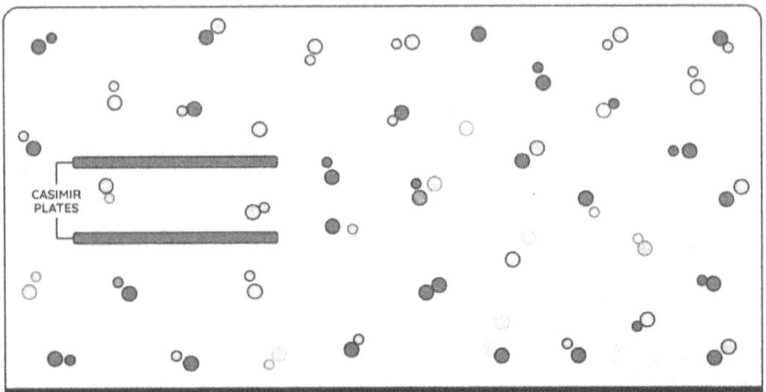

Coaching virtual particles popping into existence and annihilaling in pairs.

Theo, the retired history teacher, might find this deeply strange. He is comfortable with the idea that space can be empty. The vacuum between stars, the space inside an atom, these seem like genuine nothingness to him. But once the revelation settles in that even perfect emptiness seethes with transient particles, it opens a different view of the universe. The vacuum is not a backdrop. It is an active participant.

Virtual particles also play a role in something much closer to everyday experience. The electromagnetic force between charged particles is transmitted through virtual photons, particles of light that exist only for the briefest allowed moment. Forces in quantum mechanics are not invisible elastic bands. They are the results of a constant, blizzard-like exchange of virtual particles, made possible by the energy-time fuzziness that Heisenberg discovered.

All of this comes from the same basic trade-off that appeared in Maya's GPS circle. Pin down one quantity too precisely, and the complementary one escapes into fog.

Energy observed over a flash of time dissolves into uncertainty, and that uncertainty is enough to conjure matter from nothing briefly. The universe is, in a very literal sense, always borrowing.

The Quiet Implication: Even Empty Space Is Not Empty

The idea that space is not truly empty is one of the most philosophically unsettling conclusions of quantum mechanics. It deserves a moment of its own, because it changes the way you think about the universe at its most fundamental level.

In Newton's universe, space was exactly that. A featureless stage on which particles moved, and forces acted. The vacuum was absent. It had no content, no energy, no activity. It was the default state of nothingness that things deviated from when matter or energy was present.

Quantum mechanics replaces this picture with something fundamentally different. Because of the energy-time version of the uncertainty principle, the ground state of any physical system, including space itself, cannot have exactly zero energy. If the vacuum had exactly zero energy at every moment, that would be a perfectly precise energy measurement at every instant of time, violating the uncertainty principle. So the vacuum must have fluctuations. It must have a restless, irreducible background activity, the quantum equivalent of a perfectly still room that cannot exist because the air molecules in it will always be moving.

This background activity of the vacuum is called vacuum fluctuations or zero-point fluctuations. They are not large, and at everyday scales their effects are immeasurably small. But in the quantum world, they are everywhere, always, and they matter.

Diagram 4.9 - The Quantum Vacuum: Never Truly Empty

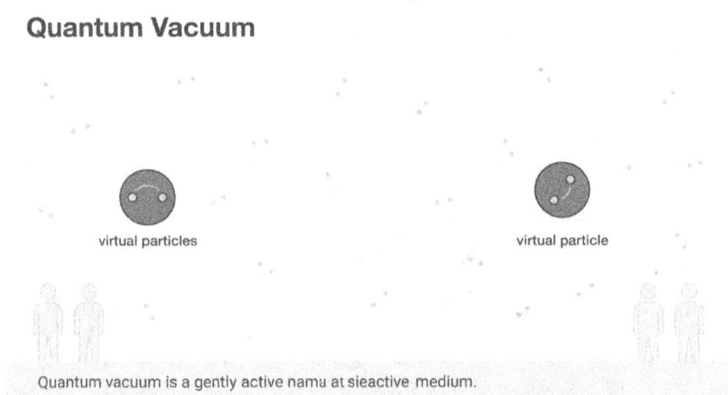

One of the most striking consequences of vacuum fluctuations involves atoms. An excited atom will eventually emit a photon and return to its ground state, even in a perfect vacuum with no incoming radiation. The vacuum fluctuations provide a tiny random trigger. This is called spontaneous emission. It is why light sources go dark when their energy supply is cut off, why a hot object's glow fades as it cools, and why fluorescent materials stop glowing when taken into darkness. Every one of these familiar phenomena traces back to the energy-time version of the uncertainty principle.

Heisenberg could not have anticipated any of this when he published his uncertainty principle in 1927. He was focused on electrons, atoms, and the immediate puzzles of quantum mechanics. But the principle he discovered was deep enough that its implications continued to radiate outward for decades, touching everything from the stability of matter to the behavior of forces to the nature of space. That is the hallmark of a genuinely great physical insight: it solves the problem you aimed at and then keeps solving problems you had not thought of yet.

What This Means For You

You will probably not be doing quantum mechanics calculations in your daily life. But the uncertainty principle shows up in the background of many things you already use, see, and rely on.

Your phone contains transistors, billions of them, in its processor chip. Transistors work by controlling the flow of electrons. In some modern transistors, the channels through which electrons flow are just a few atoms wide. At that scale, quantum effects, including the uncertainty principle, are not background noise. They are the dominant physics. Engineers who design modern computer chips must account for quantum uncertainty in their work. The miniaturization that gave you a powerful computer in your pocket was made possible only because physicists understood how electrons behave at tiny scales, including how they spread out rather than stay put when confined.

Aisha's hospital has an MRI machine. MRI stands for Magnetic Resonance Imaging. It works by manipulating the quantum states of hydrogen nuclei in the patient's body, using carefully tuned radio waves to flip those states and then detecting the energy released as the nuclei return to their ground state. The entire process depends on quantum properties, including the energy levels of particles and the ways those levels respond to measurement. Without the quantum-mechanical understanding that grew out of the work of Heisenberg and his contemporaries, MRI would not exist.

Diagram 4.10 - Quantum Uncertainty in Everyday Technology

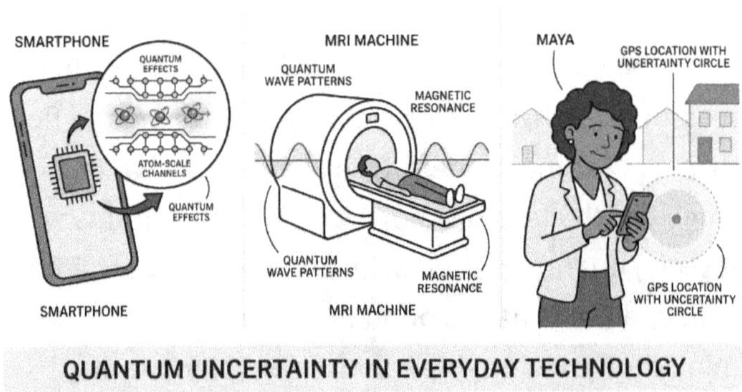

Lasers are another quantum technology that most people use every day, in barcode scanners, in fiber-optic communications, and in laser pointers. A laser works by stimulating atoms to emit light in a coordinated way. A laser's tight, coherent beam is possible only because quantum mechanics specifies atomic energy levels with remarkable precision. Those precise levels are a consequence of the wave-like, quantized nature of quantum mechanics that Heisenberg's generation discovered.

Even the solidity of the objects around you traces back to quantum uncertainty. The reason a table supports your coffee cup rather than letting it sink through is ultimately rooted in the quantum mechanical behavior of electrons in atoms. Electrons cannot be simultaneously squeezed into the same confined region, due to a related quantum principle. That refusal is what gives matter its solidity.

The sun shines because of quantum tunneling, a close relative of the uncertainty principle. The fusion reactions in the solar core require hydrogen nuclei to come extremely close together, close enough for the strong nuclear force to grab them. But the electromagnetic repulsion between positively charged nuclei should prevent that. The nuclei tunnel through the repulsion barrier, borrowing energy for just long enough for the uncertainty principle to allow it. Without quantum tunneling, the sun would not shine. Quantum mechanics lights the sky above you.

Diagram 4.11 - The Sun Shines by Quantum Rules

QUANTUM TUNNELING

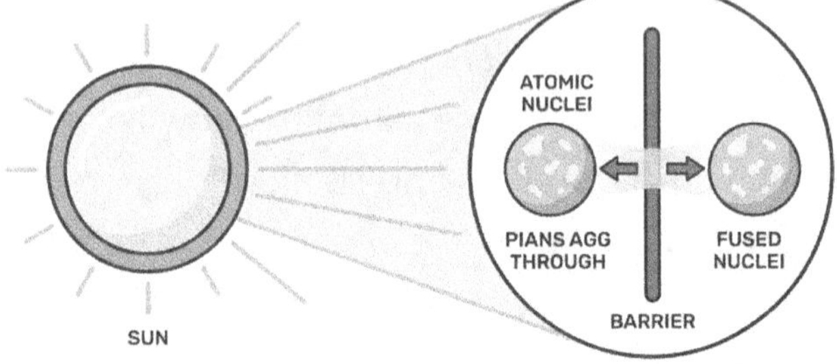

QUANTUM TUNNELING ENABLES STELLAR FUSION

When Maya finally arrives at her office and opens her laptop, she is surrounded by quantum mechanics. The screen's pixels emit light through quantum energy transitions. The chip routing her emails moves electrons through channels governed by quantum uncertainty. She never has to think about any of this. But it is all there, running quietly in the background, a technology built on the fog that Heisenberg found.

Takeaway

In everyday language, fuzziness is a defect. A fuzzy image is a bad image. A fuzzy memory is unreliable—something to be corrected, sharpened, fixed.

Heisenberg showed us that at the deepest level of reality, fuzziness is not a defect. It is a feature. Position and momentum resist being simultaneously sharp, not because our tools are crude but because they are genuinely, irreducibly paired in the structure of the universe. The fog is not something we need to clear away. The fog is physics.

This changes how you look at certainty itself. Newton's great gift was the idea that with enough information, the future could be predicted perfectly. Heisenberg's gift is more honest. He showed us that nature places fundamental limits on how much information can ever be obtained about a particle at any given moment. This is not pessimism. It is precision. It is the universe being accurate about what it actually is.

Diagram 4.12 - Newton's Sharp World vs. Heisenberg's Honest Fog

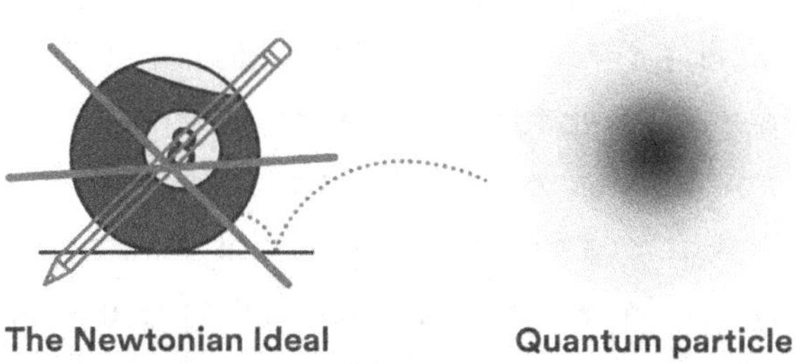

And something is empowering in this once you sit with it. The fact that electrons cannot be pinned down is the same fact that keeps atoms stable. The same fuzziness that prevents electrons from collapsing into the nucleus is the same fuzziness that holds matter together. The quantum vacuum's restless energy is what triggers spontaneous emission and fills the universe with light. The energy borrowed by virtual particles from nothing carries forces between charged particles and binds the nucleus together.

The fog is not an obstacle. The fog is what makes the universe work.

Maya's GPS circle, that soft halo of uncertainty on her screen, is a small everyday echo of all of this. Her position and her speed both carry a fuzziness that no update or improved satellite can fully remove, because at the bottom of engineering limits lies a quantum bedrock that does not yield. She will probably still find the circle slightly annoying on a rushed morning. But now she knows what it is pointing at. The circle points to the structure of reality. And reality, it turns out, was built with a certain elegant vagueness built right in.

In the next chapter, we will take this idea of quantum uncertainty one step further. We have seen that a particle's properties resist being sharply defined. Now we will see that before a measurement pins things down, a particle can be in two or more states at the same time, not just fuzzy about one property but genuinely holding multiple possibilities at once. Superposition is one of the strangest and most powerful ideas in all of physics. It is also, in its own way, exactly what you would expect from a universe that discovered long ago that vagueness can be a virtue.

5 Both at Once: The Radical Logic of Superposition

Opening Scenario

Theo was pulling weeds near the back fence when the coin fell out of his shirt pocket. It was a quarter he had been carrying for years, a habit left over from the classroom, where he used to flip it to decide which student would answer next. The coin tumbled from the pocket, hit the soft earth between two tomato plants, and settled face down in the dirt. Before he could lean forward to look, he found himself thinking about the split second before it landed. The moment when it was spinning end over end, catching the morning light—that moment when it was, well, neither one thing nor the other.

He picked it up. Heads. A perfectly ordinary result. But the thought stayed with him as he moved along the row of plants, pinching off yellow leaves. His granddaughter Celia had mentioned at dinner the week before something about quantum computers working by keeping a thing in two states at once. She had said the word superposition with the ease of a fourteen-year-old who had watched too many science videos online. Theo had nodded and changed the subject. He was a history teacher by trade, thirty years of it. He could tell you how the printing press dismantled feudal information structures. But a computer in two states at once? That sounded like nonsense.

He had also heard of Schrödinger's cat. Everyone had, sort of. A cat in a box that was alive and dead at the same time. He had always assumed this was a physicist's joke, a way of being clever at parties. But Celia had said, quite seriously, that superposition was real. That it was not just a metaphor. Theo dusted the soil from the coin and tucked it back into his pocket. Today, he was going to find out what she actually meant.

Diagram 5.1 - Theo and the Coin in the Garden

Why It Matters

Superposition is one of those ideas that tends to get flattened into a sound bite. The cat is alive and dead. The electron is in two places. The computer works on all answers simultaneously. These statements are not exactly wrong, but without context they feel like stunts. They invite the listener to shrug and move on rather than stop and ask what is actually being said.

What is actually being said is this: quantum objects, things like electrons, photons, and atoms, do not have a single definite state before you measure them. They exist in a genuine combination of possibilities, and that combination is not just our ignorance talking. It is a real physical condition with measurable, testable consequences. The consequences show up in experiments. They show up in technologies you use every day. And they are the reason quantum computers might one day solve problems that would take a classical computer longer than the age of the universe to solve.

If you really understand superposition, you will be able to see through the hype around quantum technology. You will understand why a quantum computer is not simply a faster version of the laptop on your desk. You will understand what Schrödinger was actually arguing, and why his cat was not a celebration of weirdness but a sharp protest against it. And you will have a vocabulary for one of the most fundamental features of nature, a feature that shapes the behavior of every atom in your body.

This chapter builds the idea carefully from the ground up. We start with the spinning coin, which gets us most of the way there, and then explain exactly where the coin analogy breaks down. We visit Schrödinger's cat in its proper historical context. We look at what happens to the superposition the moment you measure it. And we end with real superpositions in real devices, from atomic clocks to qubits. No math required. Just honest, step-by-step thinking.

Diagram 5.2 - The Two Ways of Not Knowing

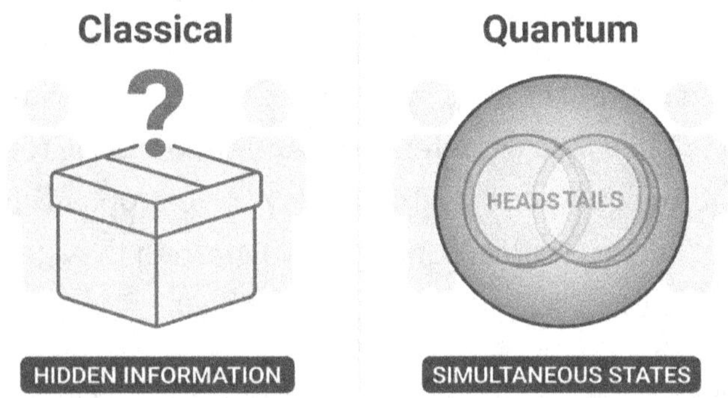

Two Kinds of Not Knowing

Here is the single most important distinction in all of quantum mechanics, and it is worth getting exactly right. There is a fundamental difference between not knowing what state something is in, and something genuinely not being in any one state at all.

Think about an ordinary coin in an ordinary box. You shake the box, set it down, and do not look inside. Is it heads or tails? You do not know. But the coin absolutely is one or the other. It has been one or the other ever since you shook the box. Your ignorance does not change the coin. The coin now has a definite face pointing upright. You cannot see it. This is classical uncertainty: the uncertainty of a missing fact. The fact exists; you have not collected it yet.

Quantum uncertainty is something entirely different. Before a quantum particle is measured, it is not hiding a secret result. There is no fact waiting to be discovered. The particle is genuinely and physically in a superposition of multiple possible states simultaneously. This is not a limitation of our instruments. It is a description of what the particle actually is. The combination is real. The particle does not make up its mind until something interacts with it in the right way: until a measurement happens.

This distinction is not philosophical hand-waving. It has been tested with extraordinary precision in laboratories around the world. The experiments are called interference experiments, and they reveal something no hidden-fact theory can explain: quantum particles in superposition interfere with themselves. They behave as though all the possibilities exist simultaneously, because all the possibilities do exist simultaneously. If the particle were hiding a definite answer, the interference pattern would vanish. It does not vanish. The superposition is real.

When Theo says the spinning coin was neither heads nor tails, he is using a loose metaphor that is mostly right and slightly wrong. The spinning coin is still a classical object. Deep in its structure, the forces of gravity and air resistance are already conspiring to determine how it will land. The outcome is hidden from us, but it is not undetermined. A quantum coin, if such a thing existed, would be genuinely undetermined until the moment it touched the earth. The two situations look identical from the outside. They are completely different on the inside.

Priya once put this to her professor: either way, you get a random result, so what is the difference? The professor said: Imagine running the same experiment a million times with the coin reset to the same initial conditions. For a classical coin, knowing all the forces, you could predict the outcome every time. The randomness is just ignorance. For a quantum particle in superposition, even knowing everything about the initial state, you still could not predict the outcome. The randomness is fundamental. It is baked into nature itself.

This is what makes quantum mechanics philosophically uncomfortable, even for physicists. It is not that we lack information. It is that the information does not exist yet. The universe is genuinely making it up as it goes along, and superposition is the state it is in while it is still deciding.

The Spinning Coin Inside an Atom

The simplest and cleanest example of quantum superposition is not a cat in a box. It is the spin of an electron.

Electrons have a property called spin. The word is borrowed from classical physics, but do not take the image too literally. Electron spin is a purely quantum property with no perfect everyday equivalent. What matters is this: when you measure the spin of an electron along any chosen axis, you always get one of exactly two results.

Physicists call them spin-up and spin-down, the quantum equivalent of heads and tails. Every single time you measure, you get one or the other. Never anything in between.

Before you measure the spin, the electron is not secretly spin-up or secretly spin-down. It is in a superposition of both. It exists in a combination of the two possibilities, and the nature of that combination matters. You can have an electron equally likely to come up spin-up or spin-down. Or you can have one weighted toward one outcome. The electron's quantum state determines the specific blend and can be any combination of the two options.

The proof that superposition is real, not just a description of our ignorance, comes from exactly this kind of experiment. Take an electron in a definite spin-up state. Measure its spin along a perpendicular axis. Now the electron is in a superposition of the new axis values. Measure the original vertical spin again. If the electron had secretly carried a definite vertical spin all along, you would get spin-up every time. You do not. The earlier measurement genuinely erased the vertical spin information because the superposition was real. Measuring it collapsed it, and the prior definite state was genuinely gone.

Diagram 5.3 - An Electron in Two Spin States at Once

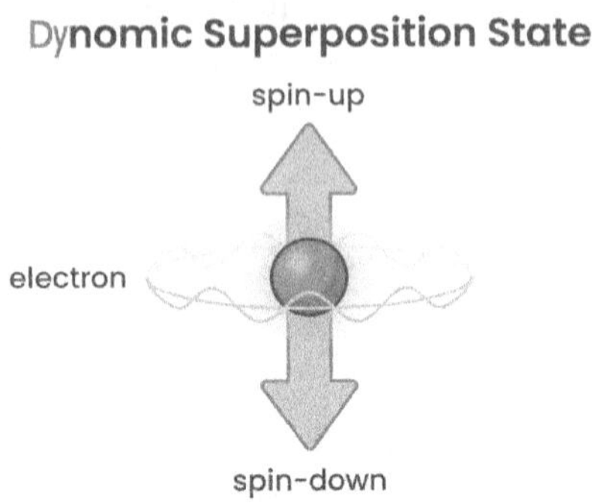

Atoms also exist in superpositions of energy levels. An atom has a ground state, the lowest-energy configuration, and excited states where the electron is higher in energy. You can place an atom in a superposition of the ground state and an excited state by hitting it with a carefully timed laser pulse. For a brief period, the atom genuinely occupies both energy levels at once. Then it settles down, emitting a photon, and the superposition ends. This is done routinely in laboratories and is the basis for atomic clocks, the most precise timekeeping devices ever built.

Think of an atom as a quantum coin that, while spinning, is literally in both heads and tails simultaneously. Not hidden, not random, but fixed, but genuinely both. The act of the coin landing is the act of measurement. The outcome becomes definite only at that moment. Before landing, both outcomes are physically present in the atom's quantum state, and experiments can detect the interference between them. That interference is the fingerprint of genuine superposition.

The energy-level superposition in atoms connects to one of the more surprising recent discoveries in biology. Plants use superposition in photosynthesis. When a photon strikes a light-harvesting molecule in a leaf, the energy must travel through a network of molecules to reach the reaction center, where it is stored. Classical physics predicts that the energy would wander through the network, losing some along the way. But experiments have shown that the energy exists in a superposition of all possible paths, exploring all routes at once and finding the most efficient one with near-perfect accuracy. Nature discovered quantum superposition billions of years before humans did, and put it to work in every green leaf.

Diagram 5.4 - Superposition in a Leaf

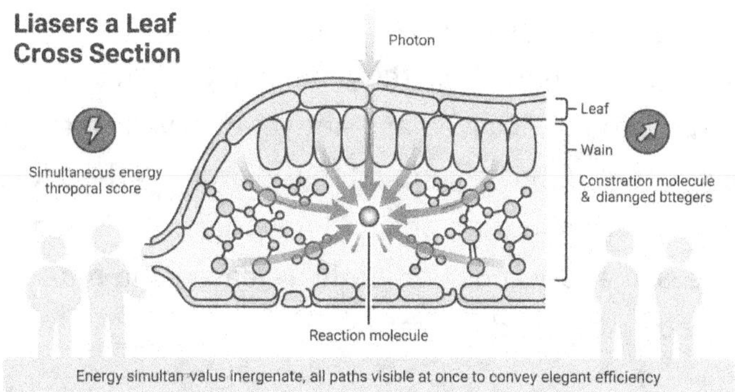

It is worth pausing to absorb what this means. Superposition is not exotic or confined to a physics building. It happens in your food before you eat it, in the chloroplasts of every plant, every second of every day. The world is quantum from the ground up. We cannot directly observe quantum behavior because our senses operate at scales where the effects average out. But they are there, working quietly, making the biology that makes us possible.

Schrödinger's Cat in Plain English

In 1935, Erwin Schrödinger described a thought experiment that has haunted physics ever since. It is important to understand from the start that he was not demonstrating quantum weirdness. He was protesting it.

Here is the setup. Imagine a sealed box. Inside the box is a cat. Also inside the box is a tiny amount of a radioactive substance, a Geiger counter to detect it, and a device that will release a lethal poison if the Geiger counter triggers. The radioactive substance has a fifty percent chance of decaying within one hour. Quantum mechanics says that before you open the box, the radioactive atom is in a superposition of decayed and not yet decayed states. The Geiger counter, being coupled to the atom, is therefore in a superposition of triggered and not-triggered. The poison device is in a superposition of the released and not-released states. And the cat...

The cat is in a superposition of alive and dead.

Schrödinger was not saying, "How wonderful, the cat is both alive and dead." He was saying, " This is absurd. The superposition principle, if taken seriously and applied all the way from quantum particles to macroscopic objects, leads to an obviously ridiculous conclusion. Cats are not in superpositions. Cats are alive, or cats are dead. Something must happen, at some point, to convert the microscopic quantum superposition into a definite macroscopic outcome. But quantum mechanics, as it was formulated at the time, had no account of when or how that happened. That gap was the measurement problem, and Schrodinger was pointing at it with deliberate, theatrical exaggeration.

Diagram 5.5 - Schrodinger's Box: The Joke That Was an Argument

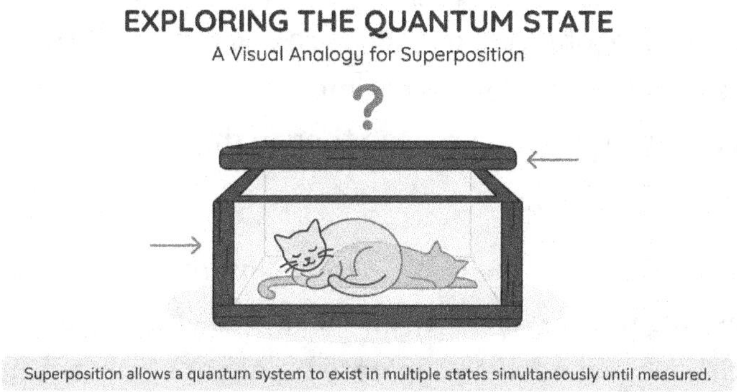

Schrodinger did not invent this thought experiment to show quantum mechanics was right. He invented it to show that it had a problem. If superposition is real and there is no rule about when it ends, the logic should extend to cats. Since that conclusion is plainly wrong by observation, either quantum mechanics is incomplete, or our account of superposition needs refinement.

Physicists have spent the better part of a century arguing about exactly this. The mainstream view, known as the

Copenhagen interpretation, holds that superposition collapses when a macroscopic device makes a measurement. This is useful but philosophically incomplete: it draws a vague line between the quantum and classical worlds without explaining where that line falls or why it exists. A more physically precise approach, called decoherence theory, explains why superpositions of large objects become practically impossible. Quantum systems constantly interact with their environment, and those interactions destroy the delicate phase relationships that make superposition possible, so quickly that no macroscopic object can maintain a coherent superposition for any detectable length of time. That is why cats are never in superpositions, even though they are made of atoms that are.

A third approach, the Many-Worlds interpretation, takes superposition so seriously that it says the cat is literally both alive and dead, in separate branches of a branching universe. This preserves the mathematics perfectly but requires an extraordinary commitment: every quantum event spawns parallel branches of reality. We will explore this in the final chapter. The key point for now is that physicists disagree not about whether superposition exists, but about what happens to it when the quantum world meets the classical one.

The cat thought experiment forces us to take superposition seriously. The quantum rules are not just a tidying-up device. They describe something real, something that demands a real explanation for why it disappears at the scale of cats and coins and gardeners. Schrödinger was right to find this troubling. Physicists are still troubled by it. But the trouble does not make superposition any less real. It means the story is not over yet.

Diagram 5.6 - Schrödinger's Cat: Two Outcomes, One Unopened Box

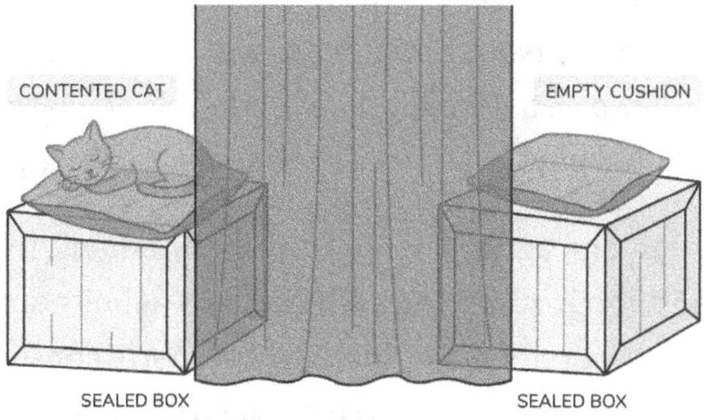

Schrodinger's Cat: Two Outcomes, One Unopened Box

What Measurement Actually Does

The word measurement sounds innocent enough. But in quantum mechanics, it is a loaded term with a specific, sometimes unsettling meaning. Understanding what measurement actually does to a quantum superposition is the key to understanding why the world looks classical even though it is quantum underneath.

When a quantum system is in superposition, its quantum state encodes the probability of each possible outcome. That state evolves smoothly and predictably over time.

The superposition grows, spreads, and interferes with itself in coherent ways. Nothing suddenly jumps. Nothing collapses. The evolution is clean and deterministic.

Then a measurement happens. The superposition appears to collapse. The quantum system goes from multiple possible states to exactly one definite state. The cat is alive or dead. The coin is heads or tails. This sudden change is called wave function collapse, though not all physicists believe a real physical collapse occurs. Some believe the other outcomes persist in other branches as well. Others treat collapse as a description of updated knowledge rather than a physical event. The debate is ongoing.

What everyone agrees on: after a measurement, the superposition is gone. The system is in a definite state. Further measurements of the same property consistently yield the same result. The coin has landed. And crucially, the act of measuring chose the outcome, not in the sense that the experimenter decided which result to get, but in the sense that the result was genuinely undecided until that moment.

Diagram 5.7 - Measurement Collapses the Superposition

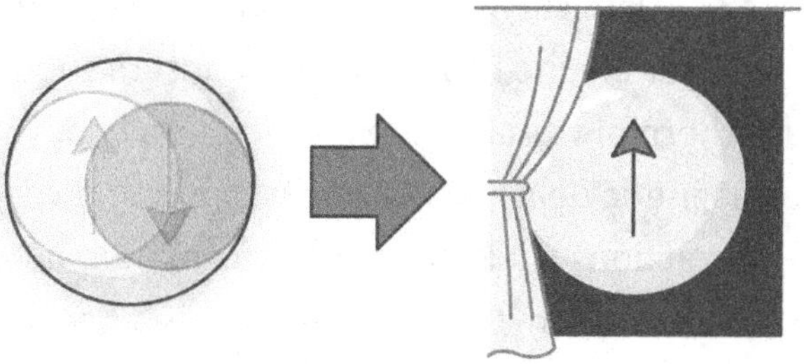

But what counts as a measurement? In the strict Copenhagen sense, a measurement happens when a quantum system interacts with a classical device large enough to produce a macroscopic, definite record: a Geiger counter clicking, a detector plate registering a spot, a screen lighting up. These interactions involve enormous numbers of particles, and their collective behavior produces the irreversible, definite outcome.

Decoherence gives a more precise account. When a quantum system interacts with its environment, whether a measuring device, a bouncing photon, or a colliding air molecule, the system becomes entangled with the environment. The quantum information spreads into an enormous number of particles and becomes impossible to recover. The superposition does not vanish mystically. It leaks into the environment so thoroughly that it might as well have vanished. For all practical purposes, the system now has a definite state. This happens extraordinarily fast for large objects but very slowly for isolated quantum systems, such as atoms in a laboratory trap.

Carlos, who spent his career diagnosing cars that behaved with perfect classical reliability, would appreciate this framing. A car engine never shows quantum superposition because thermal vibrations, air molecules, and electromagnetic fields constantly hammer it. Every one of those interactions leaks quantum information away before it can cohere into anything. The engine is a decoherence machine. So is everything we interact with in daily life. The quantum world is always there underneath, but the noise of the classical world drowns it out before we can see it.

The practical implication is this: to keep a quantum system in superposition long enough to be useful, you must isolate it ferociously from its environment. This is one of the central engineering challenges of quantum computing, and it is why quantum computers must be cooled to temperatures colder than outer space. Any amount of warmth produces vibrations that collapse the superposition within microseconds.

Real Superpositions in the Real World

Superposition is not a theoretical abstraction. It is an experimentally confirmed, technologically exploited physical reality. Here are three concrete places where real superpositions do real work in the real world.

The first is the atomic clock. Inside an atomic clock, cesium atoms are placed in a superposition of two energy states. A microwave field drives transitions between those states. The frequency at which the atoms oscillate is extraordinarily stable, one of the most stable periodic phenomena in nature. By counting those oscillations, an atomic clock keeps time accurate to within 1 second over 3.3 × 10^8 years. Every GPS satellite carries one. Every time your phone locates you, it depends on atomic superpositions to work. Superposition is why you can navigate.

Diagram 5.8 - An Atom Oscillating Between Two Energy Levels

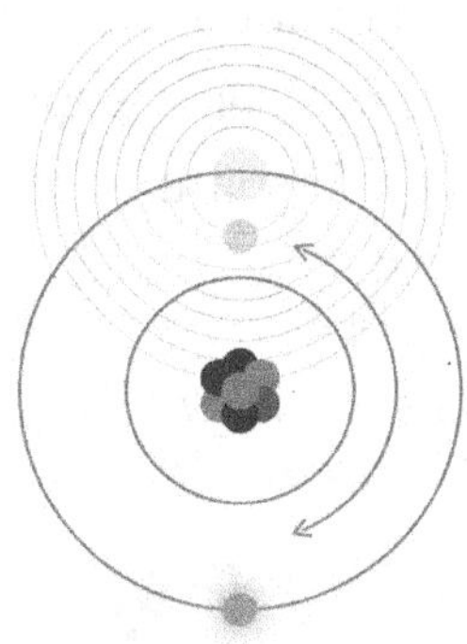

The second application is the qubit in a quantum computer. Classical computers store information in bits, each of which is either zero or one. A qubit stores information in a quantum two-state system, such as the spin of an electron, the energy level of an atom, or the polarization of a photon. Because the qubit can be in a superposition of zero and one simultaneously, a quantum computer with many qubits can process an enormous number of possible inputs at once. This is genuinely parallel processing at the quantum level, not a classical computer running calculations in rapid sequence.

It is important not to oversell this. A quantum computer is not a general-purpose speedup. It will not make your email load faster. It is particularly powerful for problems involving large-scale searches through possibility spaces or for simulating quantum systems. Cracking certain encryption schemes is one such problem. Simulating drug molecules for new medicines is another. For most everyday computing, a quantum computer offers no advantage. But for the problems it suits, it is so dramatically faster that the difference is almost incomprehensible.

Diagram 5.9 - Qubits as Spinning Coins in an Array

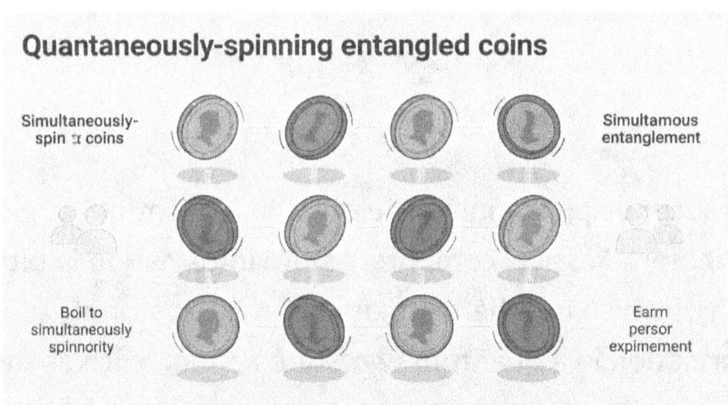

Keeping qubits in superposition long enough to compute is the central engineering challenge. Qubits decohere almost instantly when they interact with their environment. A stray photon, a vibration from a passing truck, a tiny fluctuation in the magnetic field: any of these can collapse the superposition and destroy the calculation. Quantum computers must be cooled to temperatures near absolute zero, roughly a hundredth of a degree above the coldest temperature physics allows. At those temperatures, thermal vibrations essentially stop, and the qubits can hold their superposition long enough to compute.

Diagram 5.10 - A Quantum Computer Dilution Refrigerator

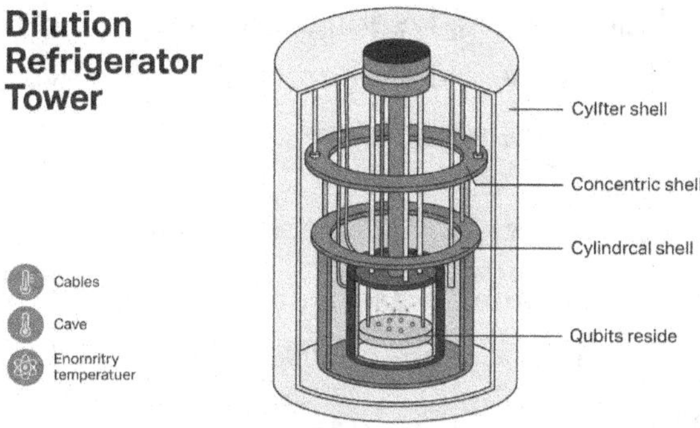

The third case is quantum superposition in biology. Beyond photosynthesis, there is evidence that migratory birds use a quantum compass based on a molecule in their eyes called cryptochrome. When light strikes cryptochrome, it creates a pair of electrons in a superposition of two spin states. The Earth's magnetic field affects the way those states evolve. The bird's nervous system detects the spin state the electrons settle into, providing directional information. The mechanism is still being studied, and some researchers debate the details.

Still, the best current evidence suggests that a bird flying south for the winter is guided, at least in part, by quantum superposition in its retina. Nature found this trick long before we did.

These three examples, atomic clocks in your GPS, qubits in a quantum computer, and the navigation of birds, span a remarkable range of scales and contexts. What unites them is superposition. The quantum ability to exist in multiple states simultaneously is not a laboratory curiosity. It is a productive feature of the physical world, harnessed by both evolution and engineering. Superposition is woven into the fabric of the world Theo moves through every day, even if most of it is invisible to him.

What It Means to Say Both at Once

We have been using the phrase "both" throughout this chapter. It is time to be more precise about what it does and does not mean.

When physicists say a particle is in superposition, they do not mean it is secretly one thing, and we do not know which. They do not mean it is rapidly flickering between two states. They do not mean it is an average of the two states. They mean something more specific: the particle's quantum state is a superposition of both possibilities, in which both are physically present and contribute to its behavior.

The evidence for this is interference. Fire an electron through a double-slit apparatus, as we explored in Chapter

3. If the electron is in a superposition of going through both slits, you see an interference pattern on the detector screen. The particle literally interferes with itself. If you try to determine which slit it went through, that measurement collapses the superposition, and the interference pattern disappears. You traded the superposition for a definite answer, and the answer cost you the interference. That is not a problem with the instruments. It is a fundamental feature of quantum mechanics.

Diagram 5.11 - Superposition and Interference

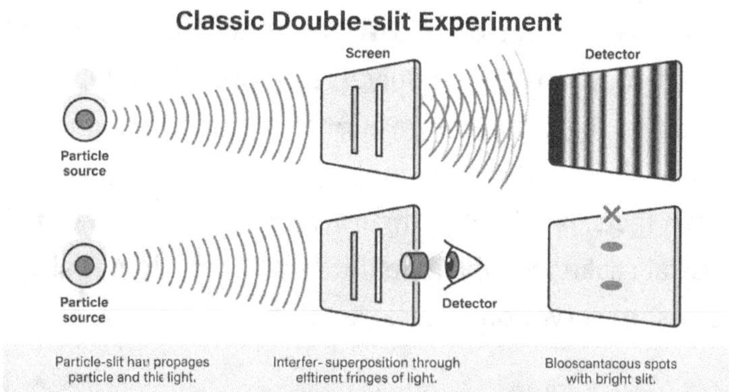

So both at once is not a poetic exaggeration. It is the most accurate plain-English description of a quantum superposition. The particle is really in both states. But it is not like being in two places in the everyday sense, where two separate copies exist. It is more like being in an indeterminate state that contains the potential for either outcome, with both possibilities actively influencing how the particle interacts with its surroundings.

Imagine a coin that genuinely has no definite face until something stops it. Not only do you not know which face it will show, but the universe has not decided. While it spins, both outcomes exist as real possibilities, and they influence each other. They interfere. They combine. Asking which face the coin has right now, mid-spin, would be asking the wrong kind of question. Only when the coin lands does the answer come into existence.

Theo's actual coin was always going to land heads. The forces acting on it from the moment it left his pocket determined the outcome, even though he could not calculate them. But if the coin were a qubit, no such predetermination would exist. The outcome would not be written anywhere in the universe before it landed. That is the radical logic of superposition. It is not about ignorance. It is not about probability in the everyday sense of not-yet-known facts. It concerns the fundamental structure of physical reality, in which definite outcomes do not always preexist observation.

What This Means For You

You might wonder whether any of this changes your daily life. In one sense, no. Cats are still alive or dead. Coins land on one face. The table is solidly there when you put your coffee cup down. Decoherence ensures that the quantum behavior of atoms averages out long before it reaches the scale of your senses. You will not be walking into any superpositions on your way to work.

But in another sense, the answer is yes, and it is a rich yes. The technology around you depends on quantum mechanics at its foundation. The transistors in your phone depend on quantum tunneling and quantized energy levels. The GPS in your car uses quantum superposition in atomic clocks. The MRI machine at Aisha's hospital uses quantum spin states to build pictures of soft tissue. The LED screen Priya uses for graphic design produces light through quantum transitions. Even the chemical bonds holding the food on Theo's table together are quantum mechanical. The classical world floats on a quantum foundation, and the engineers who built the modern world learned to take that foundation seriously.

The immediate practical payoff of understanding superposition is literacy. When you read about quantum computers, you will know what a qubit is and why it differs from an ordinary bit. You will understand why quantum computers need extreme cooling. You will not be fooled by hype that promises they will solve everything, nor will you be dismissive of what they genuinely will do. When your granddaughter mentions superposition at dinner, you will not change the subject. You will ask a follow-up question.

There is also a philosophical payoff. Our intuitions are tuned to the middle scale of reality, the scale of tables and coins and people, built by evolution for navigating a world of roughly this size. But the world does not stop at our intuitions. The very small follows different rules, and those rules are the foundation of the very large. When we let quantum systems teach us new concepts, we get atomic clocks, qubits, and an understanding of why leaves are green.

Frank once said that the most useful thing physics had done for him was to teach him that the universe did not owe him an easy explanation. It was more complicated than he thought, but in a specific and learnable way. Superposition is like that. It is genuinely strange. But the strangeness has a precise shape, and once you have that shape in your mind, the world makes a different kind of sense.

Diagram 5.12 - Theo in the Garden with a Coin and a New Understanding

Quantum wonder connects to everyday life.

Takeaway

Theo went back inside for lunch. Before he did, he flipped the coin one more time to watch it spin. This time, he paid attention to the spinning itself, the brief period before the outcome existed. His coin was classical through and through, and he knew that now. But he also knew that somewhere underneath the spinning metal, at the level of individual electrons and atomic spins, a very different kind of indeterminacy was happening, an indeterminacy that was not about ignorance but about nature.

Superposition is the quantum state in which a system exists in a genuine combination of multiple possibilities simultaneously. Not a shorthand for ignorance, not a metaphor: a physical reality confirmed by interference experiments, exploited in atomic clocks, put to work in quantum computers, and used by migrating birds to navigate continents. The superposition is real and everywhere.

Schrödinger's cat is not a demonstration of quantum weirdness. It is an argument: a deliberate absurdity designed to expose a genuine tension between the quantum and classical worlds. Decoherence gives us the best practical answer; the constant interaction of large objects with their environment destroys superposition almost instantly. But the deeper philosophical questions remain open, and that openness is a sign of honesty in science, not of failure.

Measurement collapses the superposition. Before it, all possibilities are present and interfering. After it, exactly one outcome exists. The transition between these conditions is the measurement problem, and understanding why it is a problem is itself a sign of deep understanding. You are now equipped to hold that problem clearly and to follow the argument wherever it leads.

The next chapter goes deeper into the quantum state itself, the wave function that describes not just which outcomes are possible but how probable each one is, and how that probability shifts as the system evolves. The coin is spinning. The wave function tells us everything about the spinning coin. The next chapter tells us how to read it.

6 The Ghost in the Machine: Wave Functions, Probability, and the Shape of the Possible

Opening Scenario

Carlos wiped his hands on a shop rag and stared at the screen. The hybrid SUV in bay three had pulled in an hour ago, its owner describing a faint whine that came and went, sometimes under acceleration, sometimes not. Carlos had seen this before. Hybrid drivetrains are complicated machines, and when one of them speaks up in a language of soft, intermittent sounds, they are rarely giving you a straight answer. He plugged in the diagnostic computer, let it run its scan, and watched the results populate the display. The readout did not say: the inverter is broken. It said: 70 percent chance of an inverter fault, 20 percent chance of a coolant sensor fault, 10 percent chance of other causes. The computer was not uncertain because it was broken. It was giving him the truth. The car's symptoms fit more than one profile. The data pointed in more than one direction. The machine was doing what any good mechanic eventually learns to do: working with probabilities.

Carlos has been doing this for over twenty years. He started back when cars were simpler, when the check-engine light usually meant one thing. Now the vehicles in his shop carry more software than the laptop he bought for his kids. The diagnostic readouts he gets today look less like a diagnosis and more like a weather forecast: likely this, possible that, small chance of something else. At first, that bothered him. He wanted certainty. Over the years, though, he has come to appreciate the honesty of a probability. It tells you the landscape of possibilities and lets you make the best move from there. This turns out to be, as we are about to discover, exactly how the universe works at its smallest scale.

What Carlos sees on his diagnostic screen is a snapshot of uncertainty made visible. Not ignorance, not failure, but an honest account of a system that genuinely has more than one possible state. The car is not definitely broken in one particular way. The evidence allows multiple explanations, each with a weight attached. Now take that idea, shrink it down to a scale a billion times smaller than a grain of sand, and you have arrived at the central object of quantum mechanics: the wave function. Like Carlos's diagnostic readout, the wave function does not point to a single definite answer. It lays out every possible answer, with every possibility weighted by its likelihood. It is not a statement of ignorance. It is a complete and honest description of reality at its most fundamental level.

Diagram 6.1 - Carlos at the Diagnostic Screen

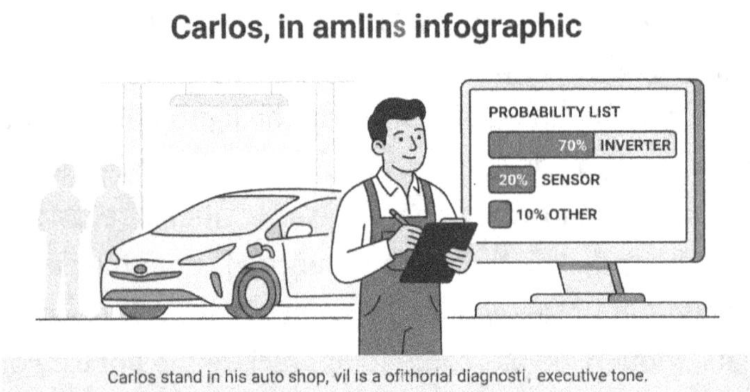

Why It Matters

The wave function might sound like an obscure technical tool that lives only inside physics textbooks. You might be tempted to file it alongside Schrödinger's cat and the uncertainty principle: interesting to read about once, then back on the shelf. But the wave function is not a curiosity. It is the engine underneath everything. Every material object that exists, every atom in your body, every electron in every circuit of every device you own, is described by a wave function. The rules of chemistry, the behavior of light, the operation of the transistors in your phone, the way DNA holds its shape, all of these phenomena follow directly from what the wave function tells us about nature.

Understanding the wave function matters because it changes how you see the world. In everyday life, we use that word loosely: it may rain; you may have left your keys in the car. We mean: we do not know for sure. But in quantum mechanics, the word possible has a sharper, stranger meaning. A particle does not simply have an unknown position; we have not yet measured it. It genuinely exists in a spread of possible positions, each with its own weight, until a measurement forces one of them to become real. The wave function is that catalog. It is not a symptom of ignorance. It is a description of how nature actually is.

By the end of this chapter, you will be able to describe the wave function to a friend without using any math. You will understand why atoms have fuzzy electron clouds rather than neat, circular orbits. You will know what the Born rule is. And you will get to sit with one of the deepest unsolved questions in physics: is the wave function a map of reality itself, or only a map of what we are allowed to know about it? Scientists are still arguing about that. You will understand the argument well enough to have an opinion.

Diagram 6.2 - The Wave Function as a Probability Landscape

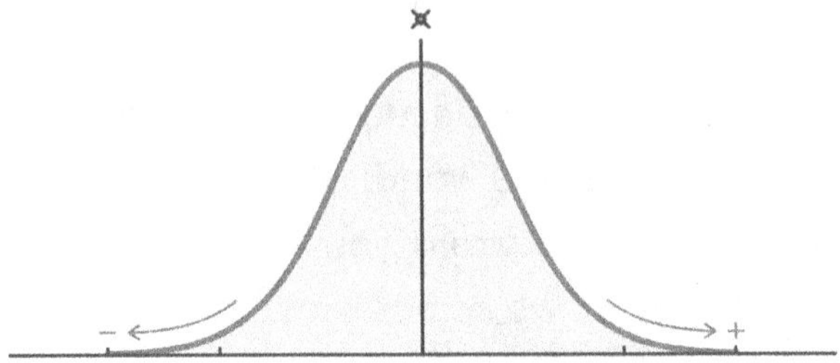

What the Wave Function Is Not

Before we can understand what the wave function actually is, we need to clear away some common images that will lead you astray. These are honest mistakes because the word function contains the word wave, and waves are things we see every day. Ocean waves roll across water. Sound waves travel through the air. Light waves ripple through space. When most people hear that particles have wave functions, they picture something similar: a tiny ripple, a physical disturbance moving through some invisible medium, a little wobble that the particle rides like a surfer.

That picture is wrong. The wave function is not a physical wave traveling through space, the way ocean waves travel through water. There is no medium it disturbs, no substance it moves through. A water wave is a real physical thing: molecules of water going up and down, passing energy from one to the next. The wave function has none of that. You cannot watch it the way you can watch the ocean. You cannot feel it, photograph it, or detect it directly. It does not live in physical space the way water or air does.

Here is what makes this important: if you carry the water-wave image into quantum mechanics, everything will confuse you. You will expect the wave function to take up space, to pass around obstacles the way water flows around rocks, and to slosh back and forth in a containable physical way. And when you read that measuring a particle causes the wave function to collapse instantly, you will ask: how can something spread across space collapse instantly? The confusion comes from the wrong starting picture. Once you let go of the physical-wave image, the collapse stops being mysterious.

Diagram 6.3 - What the Wave Function Is Not

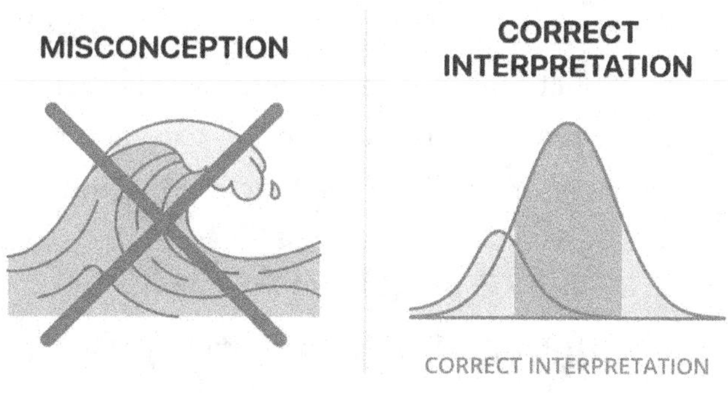

The wave function is not a path, either. Before quantum mechanics, physicists described particles as objects that moved along definite trajectories, like billiard balls following precise routes across a table. A particle started at point A, moved in a specific direction at a specific speed, and arrived at point B at a specific time. The wave function does not describe a path like that. It describes a spread of possibilities, a kind of smeared-out potentiality, not a route.

It is also not a measure of our ignorance in the ordinary sense. When Carlos looks at his diagnostic screen and sees the 70-percent reading, part of what is happening is that he lacks information. The car's real problem is already determined. It has a specific fault. Carlos does not know which one yet because he has not opened it up and looked. The probability is a symptom of his limited knowledge. Many people assume the wave function works this way, too: that the particle has a definite position, a definite momentum, a definite everything, and the wave function reflects our failure to measure it yet. This turns out to be almost certainly false. The spread of possibilities in the wave function is not about what we know. It is about what is real. That is the shocking part, and it is where the real story of the wave function begins.

What the Wave Function Actually Is (a Complete Catalog of Possibilities)

Think about Carlos's diagnostic computer one more time, but now imagine a more powerful version: a machine that does not merely reflect Carlos's limited knowledge of the car, but reflects the car's actual physical state, a state in which the inverter is genuinely 70 percent likely to be faulty, not because information is missing but because the car's condition is inherently spread across possibilities. This would be strange for a car, since cars are large everyday objects whose faults are definite even if unknown. But for a particle, this is exactly what nature provides.

The wave function is a complete catalog of all possible outcomes a quantum system can have. For a single electron, the wave function might tell you: this electron has a 60 percent chance of being found near the nucleus if you look, a 30 percent chance of being found a little further out, and a 10 percent chance of being found even further away. These are not estimates based on missing data. They are the complete description of where the electron actually is, spread across those possibilities, right now, before anyone looks. The electron is not secretly at one location, hiding from us. It genuinely occupies a superposition of all those locations simultaneously, with each location weighted by its probability.

Each entry in that catalog comes with a number called an amplitude. The amplitude is not the probability itself. It is the raw ingredient from which probability is cooked: a measure of how strongly a particular possibility features in the catalog. Some possibilities have large amplitudes and are likely to occur. Others have small amplitudes and are unlikely. The recipe for converting amplitudes into probabilities is called the Born rule, and we will get to it shortly.

Diagram 6.4 - The Wave Function as a Complete Catalog

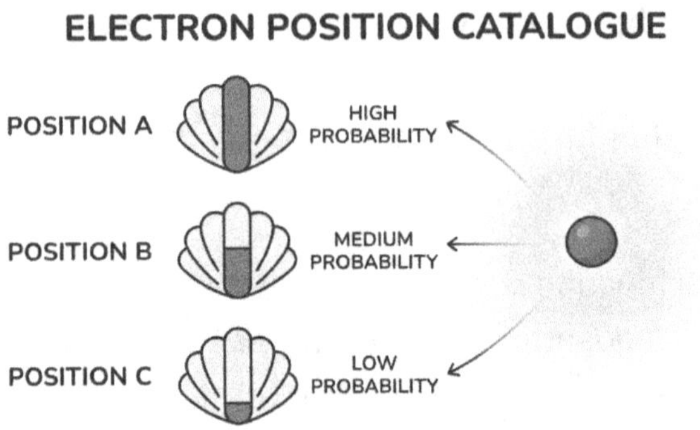

The wave function is also dynamic. It evolves according to a rule discovered by Erwin Schrödinger in 1926. Think of that rule as a time-travel machine: plug in what the wave function looks like right now, and it tells you what it will look like a moment later. The wave function flows through time in a completely predictable, smooth, continuous way. Unpredictability only enters when you make a measurement and find the particle at one specific place instead of spread across possibilities. Until then, picture it as a living catalog, constantly updating itself, telling the universe what its options are at every moment.

The catalog contains more than just position. The wave function encodes every property a particle could have: its momentum, its energy, its angular momentum, and its spin. Each property has its own spread of possibilities, weighted by its own set of amplitudes. The wave function is not just a catalog of where the particle might be. It is a catalog of everything the particle might be doing, all at once. That is why physicists call it the complete description of a quantum system: everything the particle can be is listed, weighed, and accounted for.

Reading the Wave Function Like a Weather Forecast

There is an analogy that makes the wave function feel less alien, and it is one that Carlos would recognize instantly: the weather forecast. When you check your weather app, and it says there's a 70 percent chance of rain, the app isn't telling you that rain is definitely coming. It is giving you a probability distribution over possible futures. It is saying: given everything we know about the atmosphere right now, the outcome rain has a 70 percent weight in the catalog of possible tomorrows. Other outcomes have smaller weights. The forecast is an honest, complete account of what the atmosphere is doing.

The wave function is like that forecast, but for a particle. Here is the key difference that makes quantum mechanics stranger than meteorology: in weather forecasting, the uncertainty is temporary. The atmosphere will do one definite thing, and the 70 percent is our admission that we cannot track every molecule perfectly. But for a quantum particle, the uncertainty is not a limitation of our tracking ability. The particle genuinely does not have a definite position until it is measured. The wave function is not a limited forecast made by a not-quite-good-enough computer. It is the full truth of what the particle is.

Diagram 6.5 - The Wave Function as a Weather Forecast

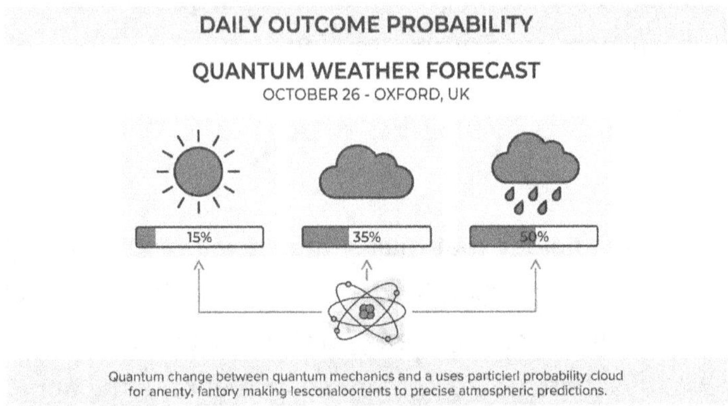

The weather analogy also helps us understand the shape of the wave function. A forecast does not give you a single number. It gives you a full picture: rain most likely in the morning, easing by afternoon, clear by evening. The probabilities are spread out over time in a particular pattern. Similarly, the wave function for a particle is spread out over space in a particular pattern. Some regions receive more probability weight, while others receive less. The overall shape tells you where the particle is most likely to be found and how spread out the possibilities are.

For an electron in a hydrogen atom, the wave function has a very specific shape. The electron is most likely to be found close to the nucleus, with probability decreasing as you move outward. The shape is roughly that of a soft sphere of probability centered on the nucleus, with density decreasing in all directions. Physicists call this shape the 1s orbital, and it looks nothing like the neat circular orbit that old textbook illustrations of atoms used to show.

Other electron wave functions have different shapes. In higher-energy states, electrons have wave functions that look like dumbbells, or hollow spheres with nodes, or complicated three-dimensional forms that bulge in specific directions. Each of these shapes is called an orbital, and each represents a different probability pattern. The wave function captures the entire probability landscape, including every region of high likelihood and every region where the particle is almost certainly not present.

This is why chemistry works the way it does. When two atoms come close together, their wave functions overlap. The probability clouds of their electrons begin to merge. When two hydrogen atoms come close enough, their electron wave functions combine to form a region of high electron probability between the two nuclei, holding them together. That shared probability cloud is what chemists call a chemical bond. It is not a stick drawn between two circles on a diagram. It is a real quantum phenomenon, a region where electron probability is concentrated because two wave functions have added their contributions. Without the wave function, there is no way to explain why any molecule holds together.

Diagram 6.6 - Atomic Orbital Probability Clouds

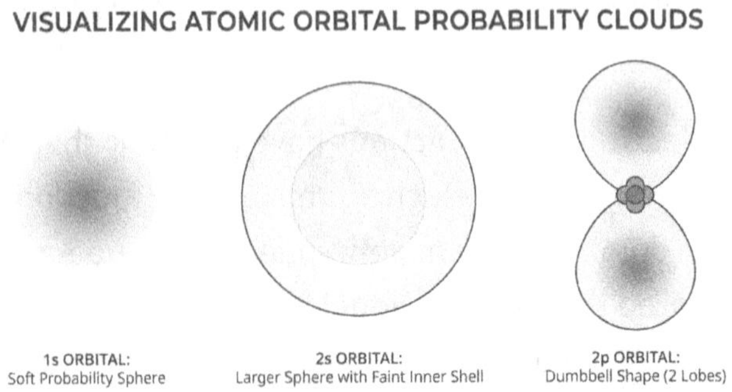

VISUALIZING ATOMIC ORBITAL PROBABILITY CLOUDS

The Born Rule: Turning Possibilities Into Odds

We have talked about amplitudes, the raw numbers associated with each possibility in the wave function catalog. Now it is time to understand how those numbers get turned into actual probabilities. The recipe for doing this is called the Born rule, named after Max Born, a German physicist who discovered it in 1926. Born's contribution was to recognize that the wave function, in its raw form, is not a probability distribution. It is something more abstract. To turn it into probabilities, you have to apply a specific operation. You have to square the amplitude.

Squaring a number means multiplying it by itself. If the amplitude is 0.5, squaring it gives 0.25, which corresponds to a 25 percent probability. If the amplitude is 0.7, you square it to get 0.49, which means roughly a 49 percent probability. The actual calculation physicists use is slightly more technical (amplitudes can be complex numbers in the mathematical sense). Still, the intuition is the same: take the raw catalog entry, apply the squaring recipe, and out comes a probability. The Born rule is the conversion table between the abstract language of the wave function and the concrete language of odds.

Diagram 6.7 - The Born Rule: From Amplitude to Probability

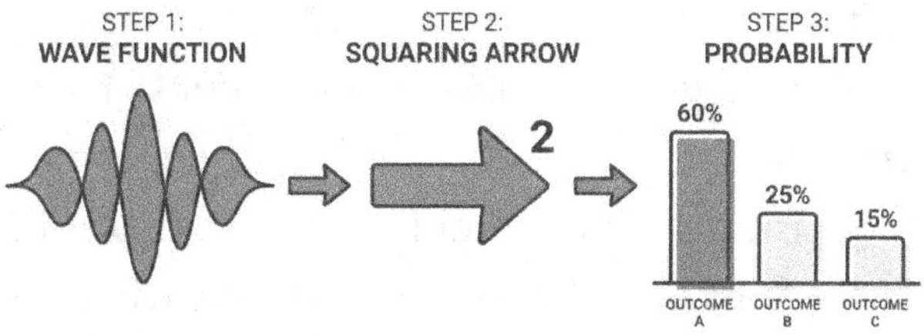

Why squaring? Why not use the amplitude directly as a probability? The honest answer is that the Born rule was not derived from deeper principles when Born first proposed it. He noticed that squaring the amplitude gave predictions that matched experimental results, and so the rule was accepted. In the nearly one hundred years since then, the Born rule has been tested against every possible experiment, and it has never been wrong. Why squaring works remains one of the deeper mysteries; some physicists have proposed derivations from more fundamental assumptions, but none have been universally accepted.

What the Born rule gives us is enormous practical power. With the wave function and the Born rule in hand, a physicist can calculate the probability of any measurement outcome. What is the chance that this electron will be found in a particular region of space? Apply Born's rule. What is the chance that a photon will pass through this filter? Apply Born's rule. It is the bridge between the abstract quantum world and the concrete world of experimental results, and it has never failed us.

Let us return to Carlos. His diagnostic computer does something analogous: it takes a set of input signals and converts them into a probability distribution. The algorithm inside the computer is its own version of the Born rule. Carlos uses those odds to decide what to investigate first. He opens the inverter housing because that is where the 70 percent lives. If he finds nothing there, he moves on to the sensor. He is working through a probability landscape, following the weights from highest to lowest, just as a physicist uses the Born rule to trace the wave function from the most probable to the least probable outcomes. The strategy is remarkably similar. The scale is just a billion trillion times smaller.

Two Camps: Real Thing or Bookkeeping Trick?

Here is where physics gets genuinely philosophical, and where even professional physicists disagree. The wave function gives us breathtakingly accurate predictions. But what is it, really? Is it a description of the world as it actually is, a real physical object out there in nature? Or is it just a bookkeeping tool, a convenient way of tracking our beliefs about a particle? These two views have names. Physicists call them the ontic and epistemic views.

Ontic comes from a Greek word meaning "existence". The ontic view holds that the wave function is real in the same way a rock or a gravitational field is real. It is out there. It exists. When an electron has a wave function spread over a region of space, that spread is not just a record of our uncertainty. It is a genuine feature of the physical world. The electron really does occupy multiple possibilities at once. Reality at the quantum level is genuinely fuzzy, genuinely spread out, genuinely indeterminate. The wave function is not a description of our knowledge. It is a description of what is.

Diagram 6.8 - Two Camps: Ontic vs. Epistemic

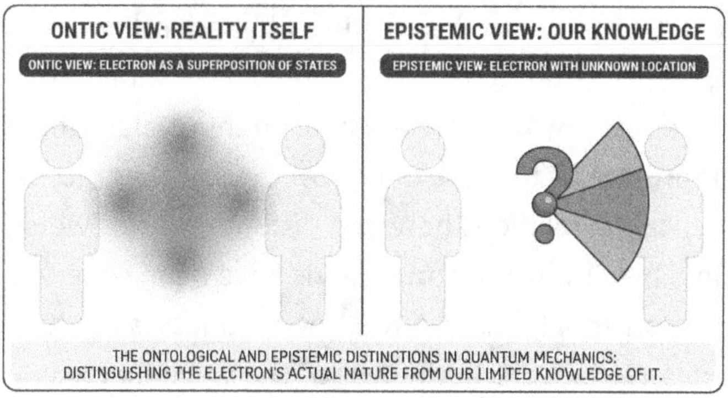

Epistemic comes from a Greek word meaning "knowledge". The epistemic view holds that the wave function records what we know, or more precisely, what we are allowed to know, about a quantum system. The particle is somewhere specific. It has a real position, a real momentum, and real values for all its properties. The wave function encodes our best possible description given the limits of quantum mechanics. On this view, the fuzziness lives in our description, not in the world.

Carlos lives in the epistemic view every day. When his screen reads 70 percent chance of inverter fault, he knows the car has a definite fault. His probability is a description of his knowledge, not of the car's actual state. The spread of probabilities is in his data, not in the engine. The epistemic view of the wave function holds that quantum mechanics works exactly as follows: the particle has a definite state. We cannot know it fully.

The problem with the epistemic view is that it runs into serious experimental trouble. A series of experiments over the past several decades (testing what are called Bell inequalities, after physicist John Bell) have shown that particles cannot have pre-existing definite values for all their properties that are unknown to us. If they did, certain statistical correlations between distant particles would be impossible. But those correlations have been observed. The data strongly suggest that the fuzziness is not just in our description. It is in the world. The wave function appears to be, at least partly, a description of how reality actually is.

And yet the ontic view has its own discomforts. If the wave function is a real physical object, it is a very strange kind of object. For a single particle, the wave function lives in three-dimensional space. But for two particles, it lives in a six-dimensional space. For a hundred particles, a three-hundred-dimensional space. This is not a figure of speech. The mathematics genuinely requires more dimensions than our physical space contains, which makes it hard to say the wave function is a straightforward physical thing sitting inside the universe we inhabit.

Most working physicists take a pragmatic position. They use the wave function as a tool, apply the Born rule, get the right answers, and move on. This approach is sometimes called the shut-up-and-calculate interpretation. It acknowledges that the theory works perfectly for prediction, while the deeper questions remain open. The debate is real and ongoing, and there is no consensus answer. What we can say with confidence is that, whether the wave function is a map of reality or of knowledge, it is the most powerful predictive tool in the history of science.

What Happens When We Measure

You have heard the phrase before: measurement changes things. But what does that actually mean? How does the act of looking affect a particle? Why does the smooth, predictable evolution of the wave function suddenly stop when we decide to check?

Before a measurement, the wave function is spread out. It contains many possibilities, each with a weight. If you are tracking an electron, the wave function might say: 60 percent chance near the nucleus, 30 percent at a middle distance, 10 percent far out. Now you make a measurement. The electron interacts with a detector. The detector registers a result: the electron is here, at this specific location. Not 60 percent here. It is here, with complete certainty, right now.

What happened to all the other possibilities? They vanished. The moment the detector registered a result, all the other entries in the probability catalog went to zero. The 30 percent, the 10 percent, and all the rest: gone. Only one outcome survived, and the wave function changed instantly from a spread-out landscape to a sharp spike at that single outcome. This sudden change is called wave function collapse. The wave function collapsed from many possibilities into a single actuality.

Diagram 6.9 - Wave Function Before and After Measurement

Wave Function Spread before and After Measurement

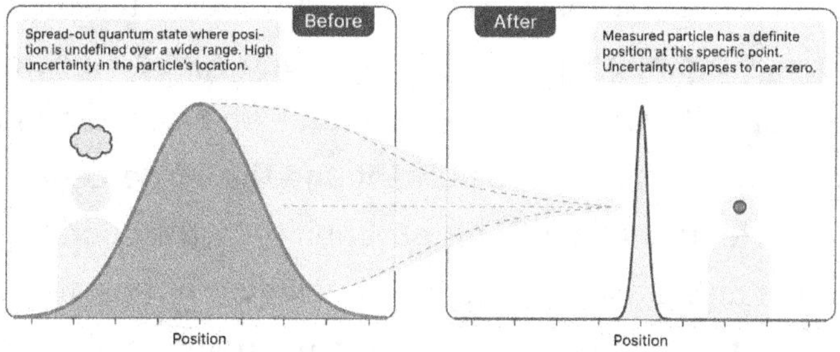

Measurement forces a quantum system from a spread-out superposition into a single defined state at one specific location.

Wave function collapse is one of the most debated events in physics. Not because it does not happen (the experimental record is clear), but because of what it means. The collapse appears to be instantaneous, meaning it does not travel through space at a finite speed. If the wave function is spread over a large region and a measurement at one edge collapses it, the rest of the wave function also collapses at the same moment, even the part far from the detector. This seems to imply reaching across space faster than light can travel.

But here is the important reassurance: no information travels faster than the speed of light during collapse. The collapse of the wave function cannot be used to send a signal. You can only know it happened after receiving information about the measurement result, and that information travels at ordinary speeds. The rules of relativity are not violated. But the strangeness remains: before the measurement, there was a genuine spread. After the measurement, there is a definite outcome. Something changed when you looked.

Think about Carlos again. He reads: 70 percent inverter, 20 percent sensor. That is his pre-measurement state. Then he opens the hood and finds a damaged wire connector. The probability instantly collapses: it is the inverter, one hundred percent. The 20 percent and the 10 percent vanish. He has collapsed his probability distribution through direct observation. The analogy is not perfect (the car's fault was always definite, just unknown). Still, the structure is exactly what happens in quantum mechanics: a spread of possibilities, an act of observation, and a definite outcome.

Diagram 6.10 - The Ghost and the Dot: Before and After Observation

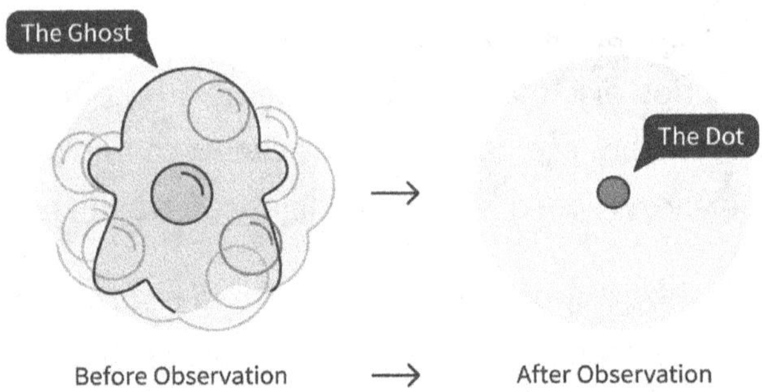

If you hold the epistemic view, collapse is not dramatic. The wave function was always just your belief catalog. When you measure, you update your beliefs. It is like finding your keys in the car: your probability of keys-in-car goes instantly to 100 percent; your probability of keys-in-kitchen drops to zero. No physical event traveled from the kitchen to the car. You just updated your knowledge.

If you hold the ontic view, collapse is much more puzzling. The wave function is a real physical object that is spread across space. The measurement event physically reduces it to a single outcome. How? What mechanism causes this? The Copenhagen interpretation (the oldest and most widely taught) says that collapse occurs when a quantum system interacts with a classical measuring device. Do not ask for a deeper mechanism. The many-worlds interpretation holds that collapse never actually happens: all possible outcomes occur in branching copies of the universe, and we follow one branch. None of the proposed interpretations has won the argument.

What This Means For You

You do not need to resolve the ontic-versus-epistemic debate to appreciate what the wave function has done for human civilization. The practical payoffs of understanding matter at the wave-function level are everywhere in your daily life.

Start with chemistry. Every reaction that happens anywhere, in a lab, in a living cell, in the ocean, is governed by the wave functions of the electrons involved. When a drug binds to a receptor in your body and relieves your headache, that binding is a quantum event: electron wave functions overlap, forming shared probability clouds that lock two molecules together in a specific geometry. The pharmaceutical industry is increasingly using quantum calculations to design molecules that bind in exactly the right way. They are engineering wave functions. They are shaping the probability landscapes of electrons to produce specific molecular behaviors.

Diagram 6.11 - The Hybrid Car and the Quantum World Inside

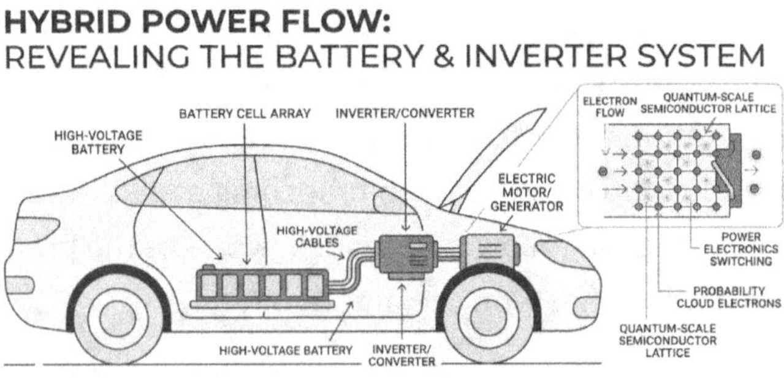

The hybrid system efficiently manages energy flow between the battery and the electric motor, with the inverter serving as a critical hub using advanced semiconductor technology.

Consider the transistors in Carlos's diagnostic computer. A modern microprocessor contains billions of them, each a tiny switch that controls the flow of electrons. Those transistors work because of quantum mechanics. Electrons in a transistor do not behave like classical billiard balls. They behave according to their wave functions. Their ability to flow, to be blocked, to tunnel through thin barriers, to respond to electric fields, all follow directly from the quantum rules we have been discussing. Without wave functions and the Born rule, the transistor would not work, and the computer would not exist.

The MRI machine in Aisha's hospital works by manipulating the wave functions of hydrogen nuclei in your body. The machine applies a precisely tuned magnetic field that interacts with the quantum spin states of hydrogen atoms in your tissues. Those spin states are quantum properties described by wave functions, and their behavior tells the machine about the density and arrangement of soft tissue in your body. The image on the screen is a translation of wave-function behavior into a picture a doctor can read.

Even the color of sunlight and the warmth of a flame are wave-function phenomena. The specific colors an atom emits when heated are determined by the allowed energy levels of its electrons, and those energy levels are the shapes of allowed wave functions inside the atom. When an electron drops from a higher-energy level to a lower-energy level, it releases a photon of a specific color. That is why neon glows red, why sodium streetlights glow yellow-orange, and why the Milky Way has a spectral fingerprint astronomers can read from billions of light-years away. The wave function is the reason the universe is colorful.

Frank might ask: Does this affect my everyday life in any way I can actually notice? The answer is yes, profoundly, but not in the way science fiction would have you believe. You do not walk around noticing wave functions. Your body is too large and too warm for quantum effects to be visible at the scale of your fingers. But the devices you use, the materials your world is made of, the chemistry of your own biology, all are quantum phenomena through and through. The wave function is not exotic. It is the foundation under everything.

Takeaway

Carlos finished his inspection, replaced the damaged inverter connector, and ran the diagnostic again. The new readout was clean: no faults detected. The probability landscape had collapsed to a single outcome: problem solved. He called the customer, explained what he had found, and quoted the repair. Simple, practical, done. He had navigated a probability distribution, followed the weights from highest to lowest, made an observation, and found certainty on the other side.

The wave function tells the same story at the very small scale. It is a complete catalog of everything a quantum system could be, with each possibility weighted by its amplitude and converted to a probability by the Born rule. It is not a physical wave traveling through space. It is not a symptom of ignorance. It is the most complete description of a quantum particle that nature allows. It evolves smoothly over time according to Schrödinger's equation, painting a shifting probability landscape until a measurement collapses it into a single definite outcome.

Whether the wave function is a map of reality or a map of knowledge is still an open question for physicists. The two camps, ontic and epistemic, are populated by serious, brilliant people with compelling arguments. The experiments so far lean toward the wave function being something more than just a record of our ignorance, but the full picture is not yet clear. This is not a failure of physics. It is physics at its most alive, wrestling with a genuine mystery, and it is one of the most honest things science does: it admits when it does not know.

What you can take away from this chapter is concrete and lasting. Atomic orbitals are not the neat circular paths in the old textbook pictures. They are three-dimensional probability clouds, the shapes of electron wave functions, and chemistry is the story of those clouds overlapping and combining. Every transistor in every device you own works because engineers learned to control electron wave functions in semiconductor materials. Every MRI scan is a direct measurement of quantum spin states. Every color of light is a wave-function transition. The quantum world is not separate from your world. It is your world, written in a smaller alphabet.

In the next chapter, we follow the wave function into its most disorienting territory. What happens when two particles share a single wave function, when their probability catalogs become permanently linked across any distance, and when measuring one of them instantly tells you something about the other? We are heading into entanglement, and it is even stranger than it sounds.

Diagram 6.12 - The Probability Landscape of Daily Life

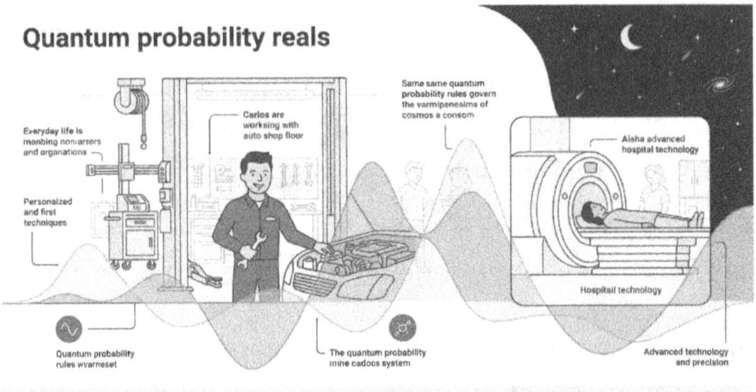

The unified quantum probability unites the unified states of datarn on these govern ruale of the cosmos.

7 Spooky Action: Entanglement and the End of Local Reality

Opening Scenario

Priya is sitting cross-legged on her dorm room bed, laptop open to a design project she keeps ignoring. Her phone lights up. She reaches for it, smiles, and types a response to her best friend Asha, who is three hundred miles away in another city. The message sends. A second later, Asha's reply comes through. They have both sent the same emoji, at what feels like the same moment, before either one could have seen what the other typed. It is a yellow cartoon face with exaggerated hearts for eyes.

Priya laughs. She takes screenshots of the exchange and sends them with a caption: "We are literally the same person." Asha replies: "Quantum twins!!" Neither of them means that literally. But Asha's joke is closer to the truth than she knows. Because inside the real world of quantum physics, there exist actual pairs of particles that are linked in a way no everyday emoji exchange can match. When you measure one of those particles, the other one instantly reflects that measurement, no matter how far apart the two particles happen to be—every single time.

This is called quantum entanglement. Albert Einstein heard about it and could not accept it. He called it "spooky action at a distance." He spent years trying to prove it was an illusion, that there had to be some ordinary hidden explanation underneath it. He was wrong. In this chapter, you will find out why he was wrong, what the experiments proved, and what this strange feature of nature actually means. One thing should be said from the start: entanglement does not let you send messages faster than the speed of light. But what it does do is strange enough on its own.

Diagram 7.1 - Priya and Asha: The Synchronized Text

CONNECTED ACROSS CITIES: SHARED MOMENTS, JUST A TAP AWAY.

Why It Matters

Most of the quantum physics in this book involves single particles behaving strangely in isolation. Quantum entanglement is something different in kind, not just in degree. It is a connection between two separate things, across any distance, that cannot be explained by anything in ordinary experience.

When physicists confirmed that entanglement is real, they closed the door on a very old and comforting idea: the idea that things far apart from each other are truly independent of each other. That idea is called local realism, and it sounds too obvious to question. Of course, the universe is built from separate, independent pieces that interact only when they physically come into contact or exchange signals.

Quantum entanglement says: no. Not always. Not for everything. Some things in the universe are connected in a way that transcends ordinary space. Measuring one changes the other, and no explanation built on local realism can account for what the experiments show. This does not mean the universe is magic. It means the universe is weirder than our intuition allows for, and experiments have forced us to accept that.

After this chapter, you will be able to explain entanglement to a friend. You will know what the EPR paradox was, what Bell's theorem proved without a single equation, what Alain Aspect's experiments showed in 1981, and what the loophole-free tests of 2015 confirmed. You will also understand why none of this lets you text Asha a secret message faster than the speed of light.

Diagram 7.2 - Two Particles, One Shared Fate

The Idea Einstein Couldn't Accept

By the early 1930s, quantum mechanics had already become one of the most successful theories in the history of science. It correctly predicted atomic behavior, light emission, and the structure of the periodic table. But it described the world in probabilities, not certainties. It said that a particle does not have a definite position or a definite speed until you measure it. Before measurement, the particle exists in a spread of possibilities. This bothered Einstein enormously. He famously said that God does not play dice, meaning he did not believe the universe was fundamentally random. He believed there had to be deeper, more precise rules underneath the fuzzy probabilistic surface of quantum theory.

In 1935, Einstein collaborated with two other physicists, Boris Podolsky and Nathan Rosen, to write a famous paper. The initials are their last names: EPR. The EPR paper was not just a complaint. It was a careful argument, and it went like this.

Imagine you create two particles together, in such a way that they share a quantum property, and then you send them flying apart. Quantum mechanics says that until you measure one of them, neither particle has a definite value for that property. They exist in a superposition of possibilities. But here is the strange part: the moment you measure one particle and discover its value, you instantly know the value of the other particle too, no matter how far away it has traveled. If the two particles are created with opposite values, then measuring the first one as "up" means the second must be "down," even if it is on the other side of the planet.

To Einstein, this seemed to have only two possible explanations. Either there was some faster-than-light influence traveling between the two particles, which he knew was impossible because nothing can travel faster than light, or the particles already had their values set from the beginning. Quantum mechanics was not telling you what those preset values were. He called these preset values hidden variables. The idea was that the particles are like a pair of gloves, each packed into a separate box. One is a left glove, and the other is a right glove. The moment you open one box and see a left glove, you instantly know the other box holds a right glove. No magic. No spooky action. The result was determined from the start, and you did not know it yet.

It was a perfectly reasonable argument. Two particles behave in correlated ways because they carry matching instructions baked in at the moment of creation. Those instructions are just hidden from view until measurement. Einstein believed this right up until he died in 1955.

Diagram 7.3 - The EPR Thought Experiment

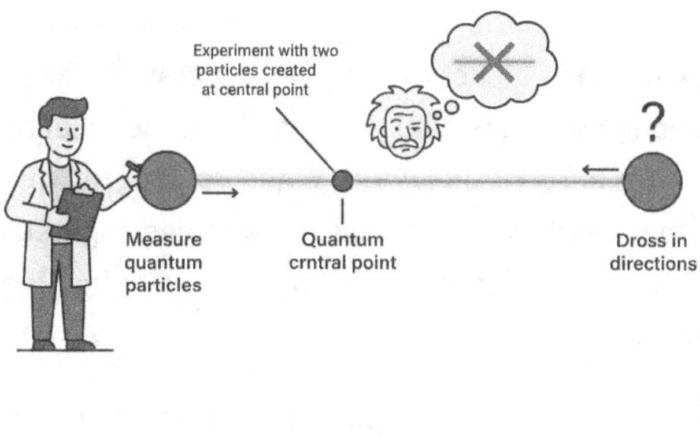

The problem was that nobody could tell the difference for twenty years after the EPR paper; the hidden-variable and quantum-mechanical ideas made the same predictions. No experiment could distinguish between them. The question seemed stuck in the realm of philosophy forever. Then, in 1964, a physicist named John Bell found the key.

The Magic Twins Metaphor

The best metaphor for entanglement is a pair of twins with a very strange property. Whenever either twin is asked a question, the answer is opposite to the answer the other twin gives at the same moment. Ask twin A whether they prefer red or blue, and they say red. Ask twin B the same question at the same moment, on the other side of the world, and they say blue—every single time. Ask them whether they prefer morning or evening. One says morning, one says evening. The answers are always opposite, always consistent, and always matched, regardless of how far apart the twins are.

Einstein's explanation for these magic twins would be that the twins made a pact before they separated. They decided in advance on all their answers, then went to their respective parts of the world. When you ask the question, each twin reads off the answer they already agreed on. No magic. No instant communication. Just planning. Hidden variables.

The quantum-mechanical picture is different. It says the twins do not decide their answers in advance. They do not carry a list of instructions. Each twin is genuinely undecided, in a real superposition of both answers, right up until the moment they are asked. But the moment one twin answers, the other twin, wherever they are, instantly shifts to the opposite answer. Not because a signal was sent. Not because of any physical connection we can detect. But because the two twins are a single quantum system that happens to occupy two locations.

In real physics, the twins are particles called photons. The question you ask them is about their polarization, which is the direction the photon is vibrating: horizontal, vertical, or any angle in between. When you create two photons together in a special crystal and send them flying in opposite directions, they share a single polarization state. Neither one has a definite polarization until you measure it. When you do measure one, the other one instantly takes the opposite polarization, no matter how far they have traveled.

Diagram 7.4 - The Magic Twins: Entangled Particles as Linked Twins

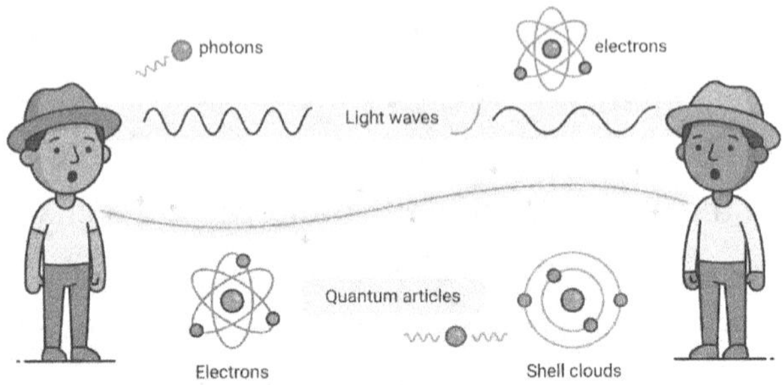

The question that Einstein and the EPR paper raised was a sharp one: how do you know the twins did not just agree in advance? How do you know the answers were not predetermined? If you measure only one property, such as polarization along a single axis, the results are perfectly consistent with the existence of hidden variables. You will always get one particle measuring horizontal and the other measuring vertical. The correlations look exactly like what you would expect from a pair of particles that were born with preset opposite polarizations.

This is where the experiment gets interesting, and this is where Bell's insight was so brilliant. If you change the angle at which you measure one of the particles, something happens that no hidden-variable theory can account for. The correlations become stronger than they are allowed to be under the rules of any world where particles carry preset answers. This is Bell's theorem, and it turns the philosophical argument into a hard experimental test.

Bell's Inequality in Plain English

John Stewart Bell was a physicist at CERN in the 1960s. He asked a question nobody had asked before: Is there a way to tell the difference between hidden variables and genuine quantum entanglement using actual measurements? His answer was yes. He expressed that answer as an inequality, which sounds mathematical but is, in fact, just a constraint.

Here is the setup. You have two entangled particles flying in opposite directions. On each side, you have a detector that can be tilted to different angles. You can measure each particle's polarization at angle A, angle B, or angle C, and you can mix and match. You run the experiment thousands of times, randomly varying the angles, and count how often the two detectors agree or disagree.

Bell's insight was this: if the particles carry hidden variables, meaning preset answers to all possible angle questions, then the pattern of agreements and disagreements must stay within a certain range. He worked out what the maximum possible agreement rate is for any world built from preset values and local influences. This is the Bell inequality. It is a ceiling. If hidden variables are real, experiments must stay below that ceiling.

Here is where it gets astonishing. Quantum mechanics predicts that the correlations between entangled particles, when you measure them at certain combinations of angles, will be stronger than the ceiling. The correlations will be too strong. They will violate the Bell inequality. They will go above the ceiling that any hidden-variable theory allows.

Diagram 7.5 - The Bell Ceiling Chart

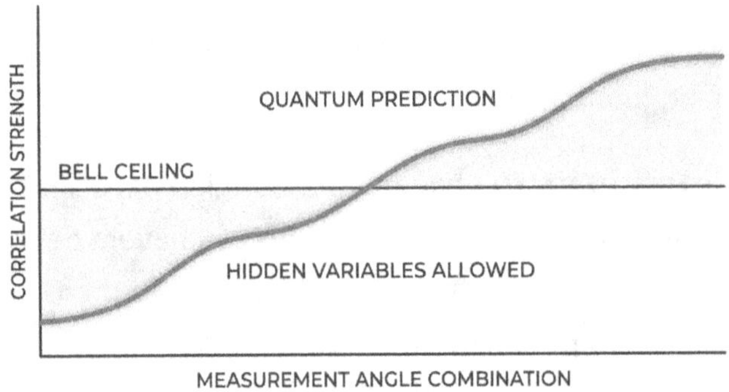

Think of it this way. Imagine you are testing whether two dice are ordinary or loaded. With ordinary dice, probability puts a ceiling on how often two dice rolled separately will both show a six. If you observe them agreeing more often than that ceiling allows, you know they are connected in some way. Bell's inequality is exactly that kind of test for particles: are these two particles following the rules of any ordinary world, or are they breaking those rules? The genius of Bell's theorem is that it turned a philosophical question into a concrete, measurable prediction. Either the world is locally realistic, or it is not. You can test it.

Bell published his theorem in 1964. He did not live to see every experimental confirmation of it; he died in 1990. But by the time he died, the answer was already very clear.

The world is not locally realistic. The particles break the ceiling. The correlations are too strong. There is no preset list of answers. The universe really is doing something that has no equivalent in ordinary experience. Bell's theorem is not a simulation or a thought experiment. It is a physical prediction that can be tested in a laboratory. When you test it, the universe answers: Einstein's preferred explanation does not work. The gloves were not packed before leaving the factory. They did not have a left and right until someone looked.

The Experiments That Settled It

Bell's theorem told physicists what to look for. Now someone had actually to build the experiment and look.

The first serious experimental tests of Bell's inequality came in the early 1970s and were encouraging. The results pointed to quantum mechanics being correct and to the existence of hidden variables being ruled out. But the experiments had technical limitations that left open what physicists called loopholes. A loophole is a way the experiment could be giving the right answer for the wrong reason.

The most important loopholes were two. The first was called the detection loophole. If the detectors only register a fraction of the photons that are sent, it is possible, in principle, that the ones being detected are not a fair sample. Maybe the undetected photons would have behaved differently. Maybe the results are skewed. The second loophole was called the locality loophole or the communication loophole. If the two detectors are close enough together that a light-speed signal could travel from one to the other during the measurement, then you cannot rule out the possibility that the two detectors are somehow influencing each other through an ordinary channel.

In 1981, a French physicist named Alain Aspect and his colleagues ran a series of experiments in Paris that were far more rigorous than anything that had come before. Aspect's team entangled pairs of photons using a carefully prepared source, sent them flying in opposite directions to two detectors, and changed the measurement angles on each side rapidly and randomly, so that no signal traveling at the speed of light could carry information from one detector to the other before the measurement was made. The results were clear and decisive: the correlations violated Bell's inequality. The violation was significant, consistent, and exactly in line with what quantum mechanics predicted.

Diagram 7.6 - Aspect's Experiment: Two Detectors, One Source

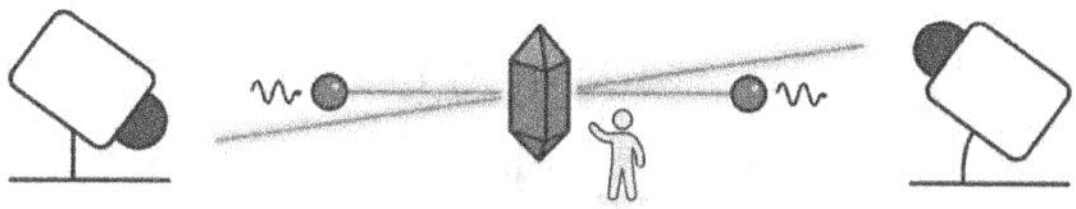

The physics community reacted with excitement and caution. Aspect's experiment convincingly closed the locality loophole, but the detection loophole remained a point of discussion. Subsequent experiments over the following decades closed the loopholes one by one. Still, it was not until 2015 that a group of scientists managed to close all the major loopholes simultaneously in a single experiment.

That 2015 experiment was conducted by a team led by Ronald Hanson at Delft University in the Netherlands. They entangled electrons rather than photons, placing the two detectors 1.3 kilometers apart inside separate buildings, with the measurements happening so rapidly and the detectors so far apart that no light-speed signal could have carried information from one side to the other. They also used sufficient detection events to simultaneously close the detection loophole. The results were clear: Bell's inequality was violated. Local realism was ruled out. This is called the loophole-free Bell test, and it stands as one of the most precisely controlled experiments in modern physics.

A second loophole-free test came from a collaboration spanning Vienna, Barcelona, and Boulder. A third major experiment that same year added further confirmation. The scientific consensus became, and remains, overwhelming: quantum entanglement is a real phenomenon. The correlations between entangled particles are stronger than any locally realistic world could produce. There is no hidden-variable story that fits the data.

Alain Aspect, along with John Clauser and Anton Zeilinger, was awarded the Nobel Prize in Physics in 2022 for their work on entanglement and Bell inequality experiments. It was a formal recognition that entanglement is not a philosophical curiosity. It is a confirmed feature of physical reality, and it has become the foundation of an entirely new field of technology.

Diagram 7.7 - Loophole-Free Bell Test: 1.3 Kilometers Apart

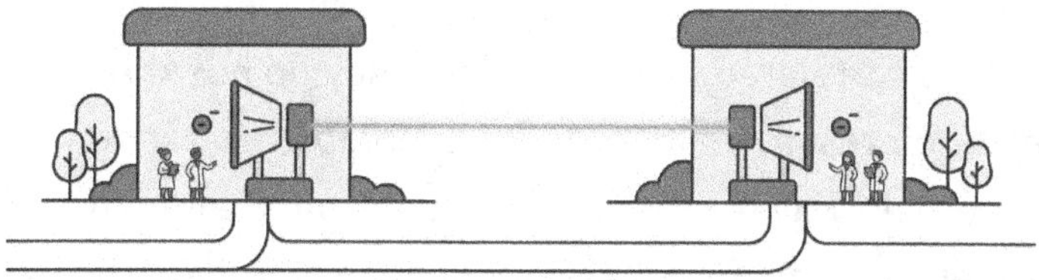

Why You Can't Send a Message Faster Than Light

This is the part that trips up almost everyone, and it is worth spending extra time on it. The correlations between entangled particles happen instantly. No experiment has ever detected any delay, even when the particles are on opposite sides of a city. Given that, you would be forgiven for thinking: why can't I use this to send a message faster than light? Why can't I encode information in the measurements on one side and have it appear instantly on the other side? The answer comes down to randomness. And the randomness is not a bug. It is the feature that enforces the cosmic speed limit.

Here is the key point. When you measure an entangled particle, you get a random result. You have no control over whether the result is up or down, horizontal or vertical, zero or one. The result is genuinely random, as far as every experiment has ever shown. You cannot choose what result you will get. What you can control is the angle at which you measure the particle, but the result you get at any given angle is still random. On the other side of the experiment, your partner also gets a random result. Their result is correlated with yours, but they do not know that yet. Their result, on its own, looks completely random. There is no way for them to tell, just from their own sequence of results, whether you have made a measurement yet or what angle you used.

The only way to reveal the correlation is to bring the two sets of results together through a classical channel, meaning an ordinary communication channel limited by the speed of light. You compare notes. You see the pattern. But the comparison itself requires ordinary communication. The instant quantum connection does not carry any readable information on its own.

Diagram 7.8 - No Faster-Than-Light Messages: Why the Channel Stays Silent

Think of it this way. Imagine Priya and Asha each receive one particle from an entangled pair. Priya measures her particle and gets the result: vertical. She knows instantly that Asha's particle is horizontal. But from Asha's perspective, she has not yet measured her particle. When she does measure it, she will get horizontal. But she has no way of knowing that Priya has already measured her particle, or what result Priya got. The result Asha sees when she measures her own particle is just a single result. It looks random. She would need to call Priya, send an email, or wait for any ordinary communication to know that their results are correlated.

That phone call or email travels at the speed of ordinary communication, which is at most the speed of light. The quantum entanglement connection is instant but invisible. It carries a correlation that can only be revealed through a classical channel. So the total communication is still limited by the speed of light. There is no loophole. There is no trick.

Physicists call this the no-communication theorem. It is not a rule invented to protect relativity from entanglement. It naturally follows from the structure of quantum mechanics. Entanglement and the prohibition on faster-than-light signaling are perfectly compatible. Popular science coverage sometimes implies that faster-than-light communication might someday be possible. That is not accurate. The no-communication theorem is not a gap that future technology will fill. Entanglement cannot carry information faster than light. Full stop.

Real-World Uses

Entanglement is not only a philosophical puzzle. It is also the foundation of a new generation of technologies. Two of those technologies are already operating in the real world.

The first is quantum key distribution (QKD). The goal of QKD is to send a secret encryption key between two parties in a way that cannot be intercepted without detection. It uses properties of quantum particles to guarantee the security of the key exchange.

Here is how it works in broad terms. Two parties, traditionally called Alice and Bob in cryptography, want to share a secret key to encrypt their messages. Using QKD, they send individual quantum particles, usually photons, between each other. The quantum properties of those photons encode the key. Because of the rules of quantum mechanics, any eavesdropper who tries to intercept and read the photons will inevitably disturb them. The disturbance is detectable. Alice and Bob can check whether anyone has been listening in on their key exchange. If the key looks undisturbed, they know it is secure. If it shows signs of tampering, they know to discard it and try again.

Entanglement can be used to further secure this process. In entanglement-based QKD, Alice and Bob each receive one particle from an entangled pair. They measure their particles independently and use the correlated random results to build a shared key. Because the correlations are random and cannot be controlled by either party, an eavesdropper cannot gain any useful information without leaving evidence in the correlations. The key is generated by nature, not by a computer, and it is provably random.

Diagram 7.9 - Quantum Key Distribution: Alice and Bob

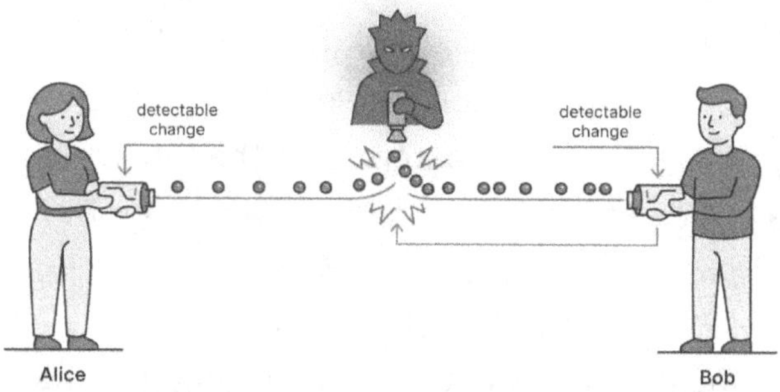

Quantum key distribution is not just theoretical. Several companies and governments have tested or deployed QKD networks over fiber-optic cables in cities. The technology works, and its security comes directly from quantum physics.

The most dramatic real-world demonstration of quantum communication came from a Chinese satellite called Micius, launched in 2016. Micius is a dedicated quantum science satellite, and it has been used to demonstrate several quantum communication protocols at distances that would be impossible to achieve over fiber-optic cables alone. Quantum signals traveling through fiber degrade over long distances, so fiber-based QKD is limited to a few hundred kilometers at most. A satellite can relay quantum signals between ground stations thousands of kilometers apart, because the photons travel through the near-vacuum of space where there is very little to disturb them.

In 2017, the team operating Micius demonstrated quantum key distribution between the satellite and ground stations in China over a distance of more than 1,200 kilometers. In 2020, they demonstrated entanglement distribution between two ground stations separated by 1,120 kilometers, with the satellite acting as a relay. The entangled photon pairs were sent from the satellite to the two ground stations, and the correlations were confirmed. The experiment showed that satellite-based quantum networks are feasible and that the principles of entanglement operate reliably even over intercontinental distances.

Diagram 7.10 - Micius Satellite: Space-Based Quantum Network

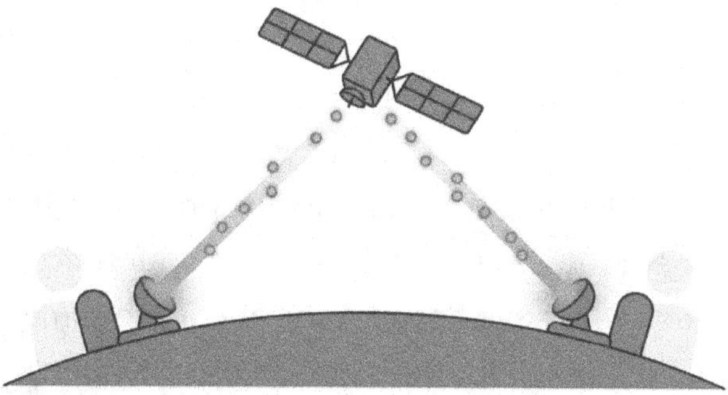

Another concept worth understanding is quantum teleportation. This term is confusing because it sounds like science fiction, in which a physical object is disassembled and reconstructed elsewhere. Quantum teleportation is not that. What it does is transfer the quantum state of one particle to another particle at a distant location, using a combination of entanglement and ordinary classical communication. The source particle itself does not travel anywhere. What moves is the quantum information, the precise description of the particle's state. The process requires a classical message to be sent through an ordinary channel, so quantum teleportation is also bound by the speed of light. It does not matter. It does not move information faster than the speed of light. The name is more dramatic than the reality.

What This Means For You

You do not need to be a physicist to feel the weight of what this chapter has covered. Let us bring it back to Priya, sitting on her dorm room bed, laughing at the synchronized emoji.

The coincidence she experienced with Asha is the kind that happens all the time between close friends. They have shared memories, shared habits, shared references. When two friends who have been talking about the same show all week both reach for the same emoji, it is not mysterious. It is a correlation through ordinary shared experience. They are not entangled particles.

But the world Priya is learning about in physics class is genuinely different. In that world, two photons created together in a crystal carry a shared quantum state. They do not carry preset instructions. They are undecided until measured. And when you measure one, the other one instantly becomes decided, with the opposite value, no matter how far away it has flown. This is not a metaphor. It is not an analogy. It is what actually happens in laboratories around the world, tested to an extraordinary degree of precision.

What does this mean for everyday life? Quantum entanglement is not yet a feature of daily experience, the way transistors are (in every device you own) or the way lasers are (in the barcode scanner at the store). Entanglement is currently a specialized tool used in research laboratories and a small number of real-world quantum communication deployments.

But that is changing. Quantum communication networks based on entanglement are being built in several countries. Quantum key distribution systems are moving from research labs into the real world of banking and government communication. The Micius satellite demonstrated that quantum communication can work over continental distances. In the coming decades, it is likely that some of the secure communications you rely on, from banking to healthcare records to private messaging, will be protected by protocols that use quantum entanglement as their security foundation.

There is also a deeper meaning here. The confirmation of Bell inequality violations changes what we are allowed to believe about reality. For centuries, the intuition that the universe is made of separate, independently existing things, that what happens here has no immediate effect on what happens there, seemed too obvious to question. Quantum entanglement shows that this intuition is wrong, at least in the quantum realm. Separate particles, created together, remain connected across space in a way that has no classical explanation.

Diagram 7.11 - Entanglement in the Real World: From Lab to Network

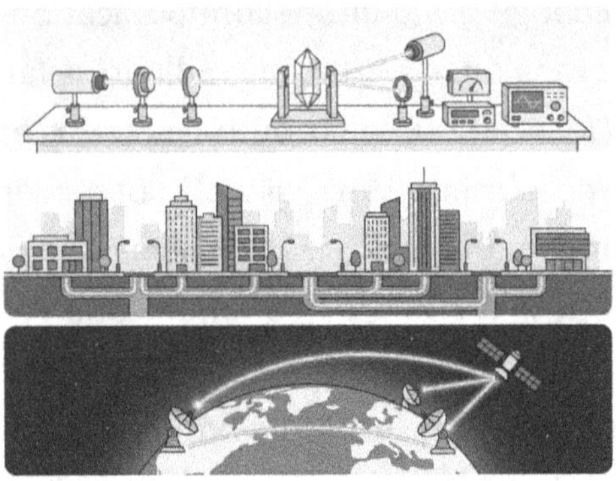

This does not mean that you and your best friend are entangled in the quantum sense, no matter how often you text the same emoji at the same time. Human-scale objects are made of trillions of particles, and their quantum properties average out into the classical behavior we experience every day. But the particles that make up you and the particles that make up everything around you do obey quantum rules at the fundamental level. The classical world you experience is built on top of a quantum foundation that is far stranger and more connected than it appears.

Knowing that is not trivial. It changes how you read headlines about quantum computing, quantum cryptography, and quantum networks. It gives you the tools to separate genuine science from hype. When someone tells you that entanglement will allow instant communication across the galaxy, you can calmly explain why that is not how it works, and you will be right. When someone tells you that entanglement is just a trick of statistics with no deeper meaning, you can explain Bell's theorem and why the correlations are too strong to be explained that way. You are now equipped to have that conversation.

Takeaway

Quantum entanglement is real. It has been tested more thoroughly than almost any other prediction in the history of physics. The correlations between entangled particles are stronger than any locally realistic world could produce. Bell's theorem proved what kind of experiment could settle the question, and Aspect's experiments, along with the loophole-free tests of 2015, settled it decisively. Einstein was wrong about hidden variables. The universe is not locally realistic.

This does not mean instant communication. The no-communication theorem is firm. The quantum correlation is real but invisible, a statistical shadow that only reveals itself when two people compare notes through ordinary, light-speed-limited channels. Entanglement is strange, but it does not break the rules that special relativity sets about the maximum speed of information.

What entanglement does is something rarer and more interesting. It shows that two particles created together can remain a single quantum system, distributed across space. Their properties are not just correlated the way the colors of matched gloves are correlated. They are genuinely undecided until the moment of measurement, and the measurement of one instantly determines the other. This is nonlocality: the system's quantum state does not belong to either location. It belongs to both.

The practical harvest of this discovery is already underway. Quantum key distribution leverages the randomness and sensitivity of quantum states to enable theoretically unbreakable encryption. The Micius satellite demonstrated that these protocols work at a continental scale. Quantum teleportation allows the transfer of quantum information across a network without physically moving particles. These technologies are not science fiction. They are physics that have been done.

And for Priya, studying graphic design in her dorm room, there is a smaller and more personal gift in all of this. The next time she and Asha send the same emoji at the same moment, she will smile and think of something real. Not because they are entangled particles. They are not. But because the universe she lives in is stranger, more connected, and more carefully structured than it has any right to be. Two photons, born together in a crystal, carry their shared fate across any distance without a wire or a signal or a whisper between them. That is not a comforting mystery or a vague inspiration. It is a confirmed fact, tested in laboratories on three continents, stamped with a Nobel Prize, and pointing toward a technology that will protect the privacy of conversations not yet held.

The universe keeps secrets well. But physics has been prying them open, one Bell test at a time.

Diagram 7.12 - The Journey from Spooky Action to Real Technology

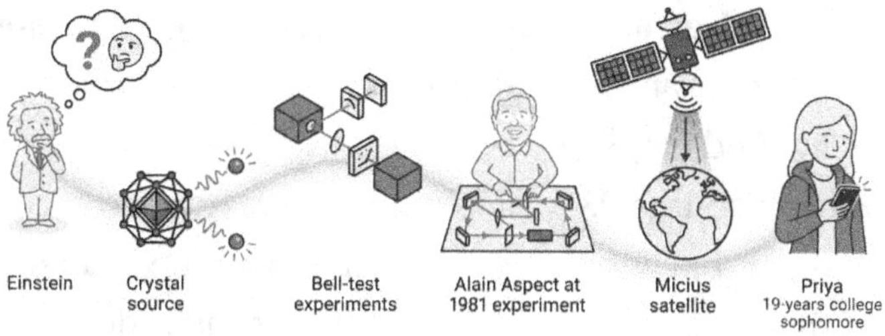

8 Through the Wall: Quantum Tunneling and the Physics of the Impossible

Opening Scenario

Aisha steps through the hospital's side door at 6:14 in the morning and feels the air shift. Inside, it smells of antiseptic and recycled warmth. Out here it smells like wet pavement and early summer, the kind of morning that makes a twelve-hour night shift feel, just for a moment, like it was worth it. She pulls her jacket tighter, tilts her head back, and watches the eastern sky go from deep grey to something almost orange. The sun is coming.

She thinks, as she often does on mornings like this, that sunlight is one of the most ordinary things in the world. It shows up every day. It warms her face on the walk to her car. It feeds every plant, drives every weather system, and powers the hospital's solar panels. Sunlight is so reliable, so taken for granted, that it barely registers as worth noticing. And yet, if you follow that warm orange light back to its source 93 million miles away, you find something that should be impossible. The sun is doing something that classical physics says it cannot do. It has been doing so for about 4.5 billion years.

Deep inside the sun's core, hydrogen nuclei are trying to fuse. That fusion releases the energy that becomes sunlight. But there is a problem. Hydrogen nuclei carry positive electric charge, and like charges repel each other. The closer two protons get, the harder they push back against each other. The energy barrier between two hydrogen nuclei is enormous. Even at core temperatures in the tens of millions of degrees, most protons do not have enough energy to overcome that barrier. Classical physics would say fusion rarely happens. And yet the sun blazes. The reason is quantum tunneling. The particles don't go over the barrier. They go through it.

Diagram 8.1 - Aisha and the Sunrise

Aisha skind outside a hospital building a friendly snit it, yet executive tone.

Why It Matters

Quantum tunneling sounds, at first hearing, like a science fiction premise. Particles passing through walls that they don't have the energy to cross. Objects appear on the other side of barriers without ever having been in the space between. It sounds like teleportation, or magic, or a mistake. But it is none of those things. It is a measurable, reproducible, precisely understood quantum phenomenon, confirmed in thousands of experiments and engineered into the technology you use every single day.

The flash memory inside your phone stores data using quantum tunneling. Every time you save a photo or a message, electrons cross a barrier that they should not be able to cross classically. The scanning tunneling microscope, one of the most precise instruments ever built, uses tunneling current to image individual atoms on a material's surface. Tunnel diodes, used in everything from microwave circuits to medical devices, are deliberately built to exploit tunneling. Radioactive decay occurs because particles within unstable nuclei tunnel through barriers that would otherwise trap them forever.

And then there is the sun. The most consequential and ancient application of quantum tunneling in the known universe is that the star at the center of our solar system is powered by it. Every photon of sunlight that has ever reached Earth, every warm morning Aisha has felt on her way to her car, every plant that has ever grown, every storm that has ever formed, traces back to quantum tunneling happening inside a star.

This chapter will show you how it works. No math required. Just a hill, a ball, and a willingness to let the universe surprise you. By the end, you will understand why a physicist can look at a sunrise and see not just beauty, but evidence that the laws of nature are stranger and more generous than common sense alone would ever predict.

Diagram 8.2 - Why Tunneling Matters

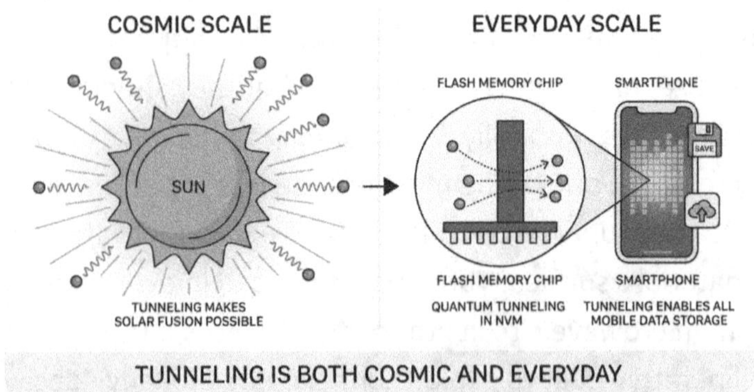

The Hill That Shouldn't Be Crossable

Start with the classical picture, because it's the intuitive one. Imagine rolling a ball up a hill. If your push was strong enough, the ball goes over the top and down the other side. If it wasn't strong enough, the ball slows down, stops on the slope, and rolls back. It never gets to the other side. This is not a subtle outcome. The ball either has enough energy to clear the peak, or it doesn't.

This is how physicists before quantum mechanics thought about energy barriers. An energy barrier is anything that requires a particle to gain energy to overcome it, such as electrical repulsion between charged particles, a magnetic field, or a chemical bond that must break. In the classical view, you compare the particle's energy to the barrier height. If the particle's energy is less, it stops and bounces back. This is the classical forbidden region. The particle is forbidden from being there, the way a ball is forbidden from the top of a hill; it can't climb.

This picture works perfectly for the everyday world. You have never thrown a ball at a wall and watched it appear on the other side. You have never seen a car roll up a hill and then vanish and reappear at the bottom on the far side. These things don't happen, and classical physics correctly explains why.

The classical picture has another feature worth noting. The height of the barrier matters absolutely. A barrier that is one unit of energy higher than the particle's kinetic energy stops the particle just as completely as a barrier that is a million units higher. From the classical standpoint, the difference between almost enough energy and nowhere near enough energy is zero. Both produce the same outcome: the particle bounces back. There is no partial crossing, no gradual leaking through. The barrier is a hard boundary.

The problem is that particles at the quantum scale don't follow this rule. And the reason comes down to the wave function. Particles at the quantum scale are not tiny balls. Wave functions, spread-out probability waves, describe them. And waves, unlike balls, do something quite particular when they encounter a barrier. They leak.

Diagram 8.3 - The Classical Barrier: Ball Stops at the Hill

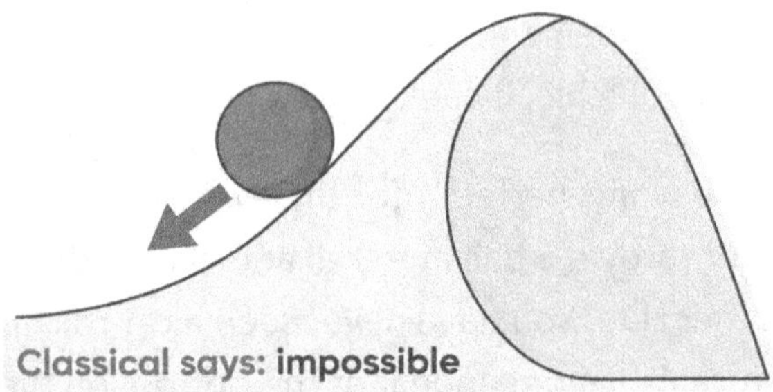

Think about a wave of water reaching a rock. The wave doesn't just stop at the rock's edge. Some of it wraps around the rock. Some get into crevices. Waves are exploratory. They don't have a single location, unlike a ball. They spread. The wave function of a particle behaves the same way when it hits a barrier.

The wave function doesn't stop at the wall. It goes into the wall. It gets smaller as it penetrates, losing strength with each step deeper into the barrier. But if the barrier is thin enough, the wave function can make it all the way through to the other side. On the far side, a smaller wave function re-emerges. That smaller wave function represents a real probability that the particle will be found there. Not a large probability, in most cases. But a real, nonzero, measurable one. That probability is tunneling.

The particle hasn't drilled a literal tunnel. It hasn't borrowed energy to climb over. The probability wave described by quantum mechanics extended into the forbidden region, and on the other side the particle had a chance of appearing. When a particle is detected beyond a barrier, it didn't have the energy to climb over it; that is, it underwent quantum tunneling.

The term "tunneling" is a metaphor, and it is worth being clear about what the metaphor is and isn't saying. The particle does not travel through the barrier as a mole does through soil. It does not pass through some hidden corridor. What happens is more subtle and more interesting. The quantum description of the particle, the wave function, has no sharp boundary at the wall. It persists on the far side. And because the wave function represents the probability of finding the particle, a nonzero wave function on the far side means a nonzero probability of detecting the particle there. When we detect it there, we call that a tunneling event.

The Wave That Leaks Through

The wave function encodes all the probabilities associated with a particle's location and other properties. Before a measurement is taken, the particle doesn't have a definite position. It has a wave function that tells us how likely we are to find it at various places. The wave function is spread out, existing in many places at once with varying degrees of probability.

Now imagine a barrier region, a region where the particle would need extra energy to exist classically. In classical physics, this region is off limits. In quantum mechanics, the wave function has no such strict boundary. It is smooth and continuous. It doesn't drop to zero at the barrier's edge. It smoothly decreases into the barrier region, still present but fading, like sound weakening as you move away from its source, but never going perfectly silent until you're very far away.

Diagram 8.4 - The Wave Function Leaking Through a Barrier

QUANTUM TUNNELING EFFECT

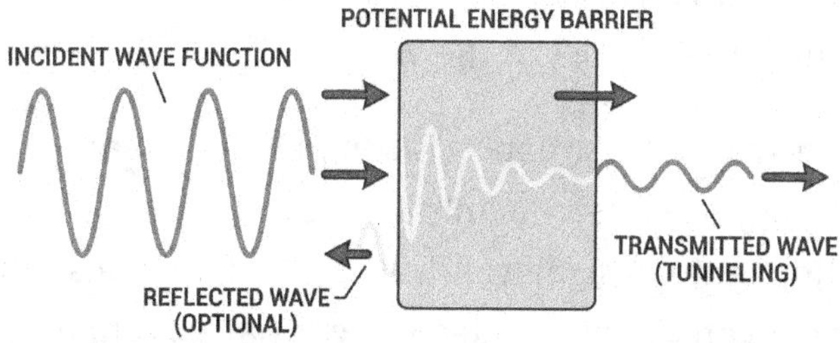

DEMONSTRATING A WAVE FUNCTION PENETRATING AND PASSING THROUGH A CLASSICALLY IMPENETRABLE BARRIER

The rate at which the wave function decays inside the barrier depends on three things. First, the height of the barrier: a higher barrier means more rapid decay, so less wave function survives to re-emerge, and the tunneling probability is smaller. Second, the width of the barrier: a thicker wall gives the wave function more distance to decay, while a thinner wall lets more survive. Third, the particle's mass: heavier particles tunnel far less effectively than lighter ones. This is why tunneling is so important for tiny particles like electrons and protons, but plays almost no role for macroscopic objects.

This answers a question you might be forming: if tunneling is real, why don't you ever see it? Why doesn't a car with slightly too little speed to crest a hill tunnel over? The answer is mass. The mass of a car is so enormous compared to an electron that the wave function of the car decays to essentially nothing almost instantly upon entering any barrier. The probability of a car tunneling through a hill is technically not zero. But it is so fantastically small that it would not happen if you waited many times longer than the age of the universe. Tunneling is a quantum phenomenon, and quantum phenomena dominate the physics of the very small.

For electrons, the situation is completely different. Electrons are extraordinarily light, and their wave functions extend significantly into barrier regions. Under the right conditions, an electron can have a tunneling probability of a few percent or even higher. That is a measurable, engineerable, useful physical probability, and engineers have been working with it for decades.

There is something philosophically important here. In classical physics, a particle's path is completely determined by its initial conditions and the forces acting on it. Either it crosses the barrier, or it doesn't, and that's calculable in advance with certainty. In quantum mechanics, even knowing the full wave function, the best you can say is that there is some probability of tunneling and some probability of reflecting. Both outcomes are genuinely possible. The universe does not decide which will happen until a measurement is made. This is not a limitation of our instruments. It is the fundamental probabilistic nature of quantum reality.

This is the same probabilistic character we encountered in Chapter 5, when we discussed superposition, and in Chapter 6, when we first met the wave function. Tunneling is not a separate exotic phenomenon bolted onto quantum mechanics as an afterthought. It is a direct, unavoidable consequence of the wave-like nature of matter. If you accept that particles have wave functions, and all the evidence says they do, then tunneling follows as naturally as the tide follows the moon.

Diagram 8.5 - Barrier Width and Tunneling Probability

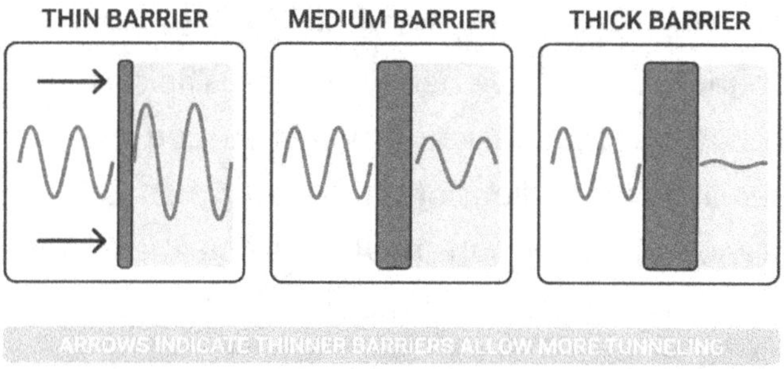

Why the Sun Shines

The light touching Aisha's face began its journey in the sun's core as a burst of energy released when two protons fused. That fusion is what powers the sun. It has powered the sun for about 4.5 billion years and will continue to do so for roughly another 5 billion. The process is called nuclear fusion, and understanding why it happens at all requires quantum tunneling.

The problem that troubled physicists for a long time: the sun's core is hot, roughly 15 million degrees Celsius, but the electrical repulsion between two protons is enormous. Each proton carries a positive charge, and positive charges repel each other with a force that grows rapidly stronger as the protons get closer. To fuse, two protons must get close enough for the strong nuclear force to grab them. But first, they have to push through the electrical repulsion barrier.

When physicists calculated the energy a proton would need to overcome this barrier classically, the number was far higher than the average energy of protons in the core. Even the fastest protons, in the high-energy tail of the distribution, didn't have nearly enough. The classical prediction was that nuclear fusion in the sun should happen at a rate millions of times lower than what we actually observe. The sun shouldn't shine at anything close to its actual brightness. Yet the sun blazes.

The resolution is quantum tunneling. The protons don't need to overcome the electric barrier. They tunnel through it. The key factors all apply: the barrier is real and high, but at the core temperatures, the protons get very close before being turned back, making the barrier they need to tunnel through quite narrow at its top. And while a proton is much heavier than an electron, it is still an extraordinarily tiny particle. Its individual tunneling probability is small, but over the immense number of proton collisions every second in the sun's enormous core, even a tiny probability, multiplied by an almost incomprehensible number of collisions, produces a steady, sustained fusion rate.

Diagram 8.6 - Proton Fusion and Tunneling in the Sun

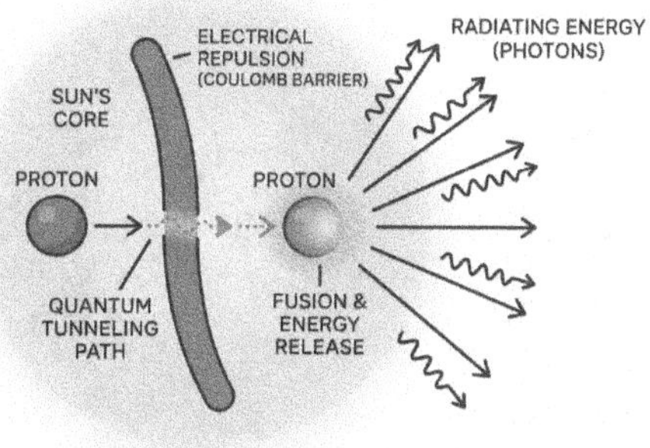

IN THE SUN'S EXTREME CORE, PROTONS USE QUANTUM TUNNELING TO OVERCOME THEIR MUTUAL ELECTRICAL REPULSION AND FUSE, RELEASING IMMENSE ENERGY.

The physicist George Gamow worked out the mathematics of tunneling in nuclear physics in the late 1920s, connecting quantum mechanics to the grandest objects in the universe. Stars are quantum engines. They run on quantum rules. Without quantum tunneling, stars as we know them could not exist. The sun would not produce enough energy through fusion to maintain its current brightness. Life on Earth would not have evolved as it has.

There is something genuinely striking about this. When Aisha stands outside her hospital and feels sunlight on her face, she is experiencing the output of a process governed by the same quantum mechanics that underlies her hospital's MRI machine, the flash memory in her phone, and the radioactive tracers used in medical imaging. The quantum world is not somewhere else. It is everywhere, all the time, including 93 million miles above the horizon.

Diagram 8.7 - The Sun as a Quantum Engine

TUNNELING PROCESS

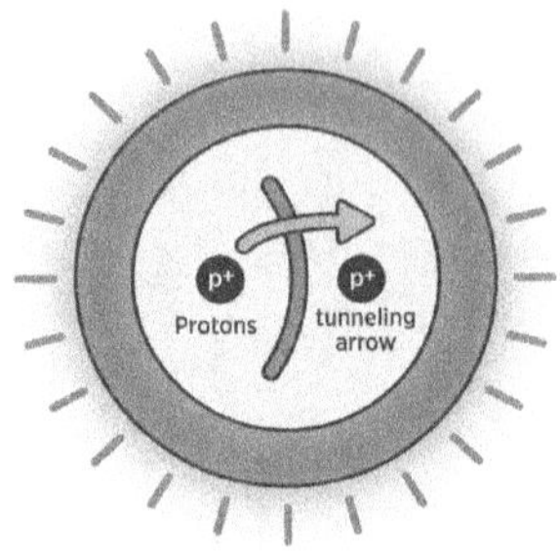

STYLIZED SUN SHOWN AS A LARGE GLOWING CIRCLE IN BLUE PALETTE WITH ENERGY RADIATING OUTWARD. INSIDE THE CORE, A SMALL INSET SHOWS THE TUNNELING PROCESS: TWO TINY CIRCLES REPRESENTING PROTONS WITH A CURVED BARRIER BETWEEN THEM AND A TUNNELING ARROW.

Radioactive Decay in Plain English

Tunneling doesn't only occur for particles trying to enter a nucleus. It also happens to particles trying to get out. This is the physics behind radioactive decay.

Some atomic nuclei are unstable. They have too many protons, too many neutrons, or their internal arrangement carries more energy than a lower-energy arrangement would. These nuclei release that excess energy by emitting a particle or a burst of radiation. One of the most important types of tunneling to understand is alpha decay, in which a nucleus ejects a package of two protons and two neutrons called an alpha particle.

An alpha particle inside a large unstable nucleus is in a peculiar situation. Inside the nucleus, the strong nuclear force creates a kind of cage. The alpha particle bounces around inside this cage. But outside the nucleus, there is an electrical repulsion barrier. If the alpha particle could get outside, it would be repelled by the remaining nucleus and fly away. The question is: how does it get from inside the cage to outside? In classical physics, it can't. The electrical repulsion barrier is higher than the energy of the alpha particle. It should stay trapped forever. And yet, on measurable timescales, unstable nuclei do decay. The mechanism is quantum tunneling.

The alpha particle is a quantum object described by a wave function. That wave function leaks into the barrier region outside the nucleus. If the wave function makes it through, there is a probability of finding the alpha particle on the outside, where electrical repulsion would propel it away. When that happens, the decay occurs. The nucleus loses two protons and two neutrons and transforms into a different element.

Diagram 8.8 - Alpha Particle Escaping a Nucleus by Tunneling

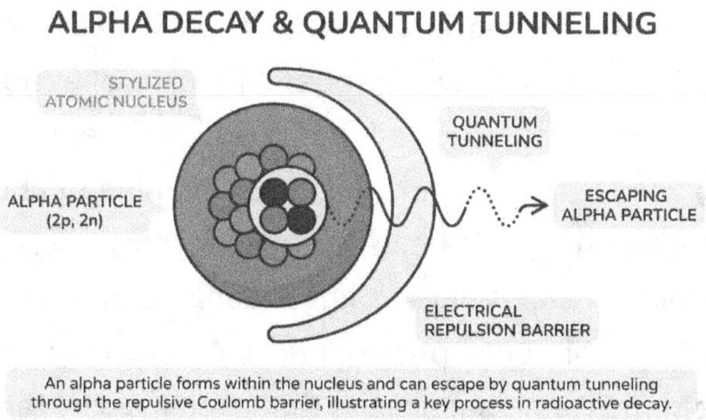

George Gamow worked out the tunneling interpretation of alpha decay in 1928, independently and simultaneously with Ronald Gurney and Edward Condon. This explained something that had puzzled physicists for years: why do different radioactive isotopes have such wildly different half-lives? The half-life of a radioactive element is the time it takes for half of a sample to decay. Some isotopes decay in fractions of a second. Others take billions of years.

The answer lies in the barrier geometry. Small differences in barrier height and width produce enormous differences in tunneling probability. An alpha particle facing a slightly thicker or taller barrier has a much smaller probability of tunneling on any given bounce. If the probability per bounce is smaller by a factor of a thousand, the average escape time is a thousand times longer, and the half-life is a thousand times greater. The enormous range of half-lives observed in nature is directly attributable to the sensitivity of the tunneling probability to barrier parameters.

This has practical consequences that reach into Aisha's working life. Radioactive isotopes with carefully chosen half-lives are used in nuclear medicine. Technetium-99m, widely used in medical imaging, has a half-life of about six hours, long enough to be useful in a diagnostic scan but short enough that it doesn't linger dangerously in the body. The entire enterprise of nuclear medicine depends on the same tunneling physics that explains why the sun shines.

The Microscope That Sees Atoms

Tunneling happens naturally in stars and radioactive nuclei. In the second half of the twentieth century, engineers and physicists learned to exploit it deliberately, building instruments of remarkable precision. The first and most dramatic example is the scanning tunneling microscope, or STM.

The STM was invented in 1981 by Gerd Binnig and Heinrich Rohrer at IBM's research laboratory in Zurich, Switzerland. Binnig and Rohrer received the Nobel Prize in Physics just five years later, in 1986. What the STM can do is almost incomprehensible from an everyday standpoint: it can image individual atoms on the surface of a material, one by one, with enough precision to distinguish where one atom ends and the next begins.

The STM works entirely on quantum tunneling. The microscope has a metal tip, sharpened to a point, ideally just one atom wide at its end. This tip is brought extremely close to the surface being examined, close enough that the gap between tip and surface is only a nanometer or two, roughly the diameter of ten hydrogen atoms side by side. At this distance, the wave functions of electrons in the tip and electrons on the surface overlap slightly, even though there is a vacuum gap between them.

Diagram 8.9 - Scanning Tunneling Microscope Tip Over Atoms

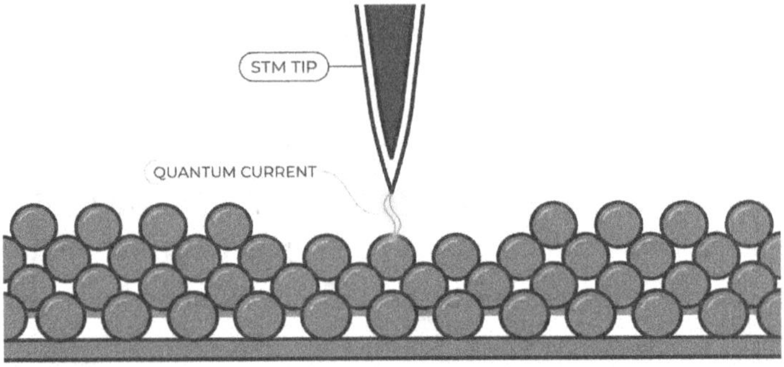

QUANTUM CURRENT ILLUSTRATOR

Because the wave functions overlap, electrons can tunnel across the vacuum gap, producing a tiny but measurable tunneling current. The key insight is that this current is extraordinarily sensitive to the gap distance. Even a change of a fraction of a nanometer produces a large and easily measurable change in current. This sensitivity is what makes the STM a powerful imaging tool.

In imaging mode, the STM moves its tip across the surface in a grid pattern while continuously adjusting its height to maintain a constant tunneling current. Wherever there is an atom, the tip rises slightly. Where there is a gap between atoms, the tip dips. By recording the height adjustments as a function of position, the STM builds up a topographic map of the surface, atom by atom. The resulting image has a resolution that no optical microscope can approach, because optical microscopes are limited by the wavelength of light, which is far larger than an atom.

In a famous 1989 demonstration, IBM scientists used an STM to spell the letters I, B, and M by moving individual xenon atoms on a nickel surface. It was a spectacular proof that humans could operate at the atomic scale with precision. Today, STMs are standard tools in materials science, nanotechnology, and chemistry. They are used to probe surfaces atom by atom, test theoretical predictions about electron behavior, and prototype molecular-scale devices. The STM is a machine that makes the quantum world visible, producing images directly from the quantum tunneling of real electrons across a real nanometer-scale gap.

Diagram 8.10 - STM Image of Atoms on a Surface

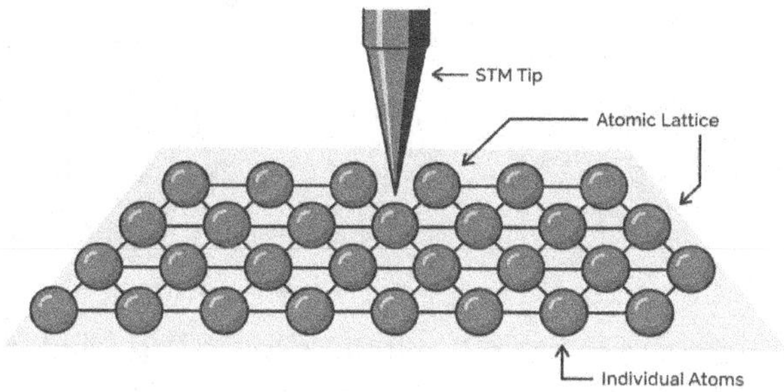

A fop-one strong metaphos: in of STM generated image concept.

Tunneling in Your Pocket

The STM lives in laboratories. For most people, the most intimate encounter with engineered quantum tunneling happens in the pocket. Flash memory, the storage used in phones, laptops, solid-state drives, digital cameras, and USB drives, uses quantum tunneling as a fundamental part of its operation. Not as a side effect. Not as a problem to engineer around. As the mechanism.

Flash memory stores information in tiny components called floating-gate transistors. Each transistor is a small structure that can trap or release electrons on an isolated layer called the floating gate. When electrons are trapped there, the transistor reads as a one. When the gate is empty, it reads as a zero. The string of ones and zeros stored across millions of these transistors is the digital data that makes up your photos, messages, music, and apps.

To write data to a flash memory cell, electrons must be injected into the floating gate. The floating gate is surrounded by a thin layer of insulating material, typically silicon dioxide, which is expected to prevent electrons from entering or leaving the gate. It is an energy barrier. The electrons don't have enough energy to climb over it. But the insulating layer is made very thin, thin enough that electrons can tunnel through it when a voltage is applied. To erase data, a different voltage coaxes the electrons to tunnel back out. The data persists because the floating gate holds its electrons firmly once they tunnel in.

Diagram 8.11 - Flash Memory and Quantum Tunneling

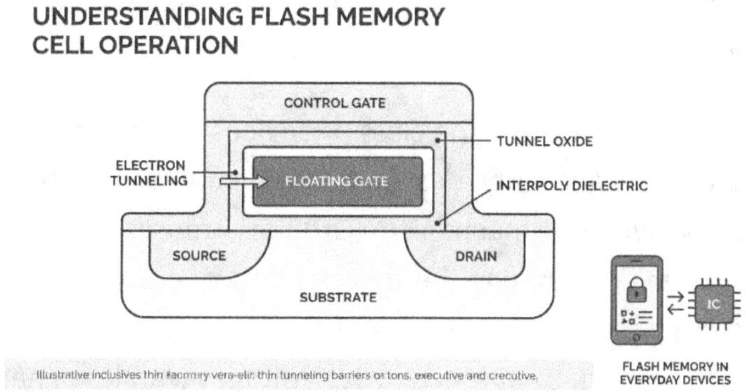

The quality and reliability of flash memory, how long it lasts, how many read/write cycles it can handle, and how quickly it stores data are all directly traceable to the physics of tunneling through that thin insulating layer. Materials scientists and electrical engineers carefully optimize the thickness and composition of the tunneling layer to balance write speed, erase speed, retention time, and durability. Every time a phone manufacturer advertises faster storage speeds, they are advertising an improved understanding of quantum tunneling.

Before flash memory, there were tunnel diodes, invented in 1958 by Leo Esaki, who received the Nobel Prize for the work in 1973. A tunnel diode works differently from ordinary diodes because, at certain voltages, electrons tunnel through a barrier in the semiconductor junction rather than flowing over it. This gives tunnel diodes unique properties: they operate at extremely high speeds (because tunneling is essentially instantaneous) and function at very low voltages. Tunnel diodes still find applications in specialized electronics, including microwave circuits, radar systems, and certain medical equipment.

There is also an emerging class of devices called tunnel field-effect transistors, designed to use tunneling as the primary switching mechanism. Ordinary transistors switch by allowing or blocking the thermal flow of electrons over an energy barrier. These newer designs use tunneling to switch, potentially at much lower voltages with less energy dissipated per switch. As conventional transistors shrink toward the physical limits imposed by atomic scales, tunneling-based devices are seen by many researchers as a promising path forward for more powerful and more efficient computing hardware.

What This Means For You

Quantum tunneling sounds remote and abstract at first hearing. But it is threaded through your everyday life in ways that are both practical and profound. The more you know about it, the more you see it.

Start with your devices. The phone, the laptop, the digital camera, the USB drive, the solid-state hard drive: all store data using flash memory, which runs on tunneling. If you use any device that records or stores digital information, you are using tunneling technology every day. Not as a metaphor. Literally, at the level of electrons moving through nanometer-thick barriers every time your device writes a file.

Medical technology is another area where tunneling touches everyday life. Radioactive tracers used in certain medical imaging techniques are chosen partly based on their half-lives, which are governed by tunneling physics. The nuclear medicine procedures some of Aisha's patients require depend on radioactive isotopes whose decay rates are, at a fundamental level, tunneling probabilities. Tunnel diodes are also used in some medical electronic devices.

There is also a more fundamental way tunneling touches your life. The chemistry inside every living cell involves reactions between molecules and atoms. Many of these reactions, particularly certain enzymatic reactions, are influenced by quantum tunneling. Protons and hydrogen atoms, being very light, can tunnel through chemical reaction barriers, thereby affecting reaction rates. Some researchers believe tunneling plays a role in enzymes involved in DNA replication and repair. The boundaries between quantum physics and biology are not as sharp as they once seemed.

What unifies all of these examples, from the sun to the STM to the flash chip to the cell, is the same underlying physics. Particles are not little bullets. Probability waves describe them. Those waves don't respect classical boundaries. They probe every available path, including through barriers that classical logic says should be impassable. When we build instruments or devices that take advantage of this, we call it engineering. When it happens naturally in stars, nuclei, and cells, we call it physics. But it's all the same thing. It's the universe following its own rules, which turn out to be quantum mechanical all the way down.

Here is a way to think about it that might help it stick. You have used a locked door. You know that if it is locked, you can't walk through it. You have to open it first. That is the classical picture: the barrier stops you completely. Quantum tunneling is nature's way of allowing a small but real probability of passing through a locked door without opening it. For sufficiently small and light particles, the locked-door rule doesn't apply with perfect rigidity. The barrier is real, but it is not absolute.

The Sun has been walking through a locked door for 4.5 billion years. Every second, vast numbers of proton pairs try to fuse in the solar core. Most fail. But a small fraction tunnels through the electrical barrier and fuses, releasing energy that eventually reaches Earth as sunlight. Without that small, steady tunneling probability, the sun would not shine at its current brightness. The sky outside Aisha's hospital would be far darker, and the world would be an entirely different and much less hospitable place.

Takeaway

Aisha walks to her car in the early morning light. The sun is fully above the horizon now, the sky gone from orange to a clear pale blue. She unlocks the car, tosses her bag in the back seat, and sits for a moment before starting the engine. Her phone buzzes on the seat beside her. A text from her sister. She picks it up and reads it.

She doesn't think about the electrons that store that message by tunneling through nanometer-thin barriers. She doesn't think about the protons tunneling through electrical barriers deep inside the sun right now, or the radioactive isotopes in the nuclear medicine department one floor below, where she worked all night. These things are just the world. They always were.

But here is what you now know that you didn't know before. The ordinary world runs on extraordinary physics. The rule that says a particle can't cross a barrier it doesn't have enough energy to cross is a classical rule. And the universe is not classical. It is quantum. In the quantum world, barriers are not absolute walls. They are probabilities. They are hills with a small but real chance of passage for any particle small and light enough to exploit the wave-like nature of matter.

Quantum tunneling is not a loophole or a glitch in the universe's programming. It is a feature. It is written into the fabric of how matter behaves at its smallest scales. It reaches all the way up, through billions of years of stellar burning, through the evolution of life on an energized planet, through the invention of transistors, microscopes, and memory chips, all the way to the phone buzzing in Aisha's hand on a Tuesday morning in early summer.

The sun shines because particles tunnel. Your phone remembers because electrons tunnel. Atoms decay because particles tunnel out of cages that should hold them forever. You now see all of this. The world has not changed. But the way you see it has.

In the next chapter, we pull back and look at the full scope of what quantum physics has given us in technology: lasers, transistors, MRI machines, and the computers that may one day think in superpositions. Tunneling will be there, woven through all of it, because the quantum world does not compartmentalize. It runs through everything.

Aisha starts the engine and pulls out of the parking lot. The sky is bright now, fully morning. The street is filling with the first signs of the day: a delivery truck, a jogger, and a traffic light turning green. All of it is ordinary. All of it, if you follow the physics far enough down, is quantum. She turns onto the highway and drives home through a world that runs on rules she now understands a little better than she did yesterday.

Diagram 8.12 - Aisha and the Quantum Morning

Aisha, ER scrubs

9 The Quantum World at Work: Lasers, Transistors, MRI, and the Computer That Thinks in Superpositions

Opening Scenario

Aisha clips the patient chart outside Room 7 and takes a slow breath. It is 6:40 in the morning. A 52-year-old man presented overnight with a dull lower backache that had been worsening over the past 3 weeks. The ER doctor suspects something near the kidney. She wants a clearer picture before deciding anything. So Aisha walks the patient to the MRI suite at the end of the hall, settles him onto the narrow table, and slides him slowly into the wide white tunnel of the scanner.

Patients almost always ask her how it works. She gives the honest answer: the machine uses the hydrogen atoms inside your body as tiny radio transmitters. The magnetic field nudges those atoms into a particular spin state, then the machine listens as they relax and transmit back. A computer turns those signals into a picture. She watches the patient's anxiety shift toward fascination. The machine is not frightening once you understand it. And every step of it depends on quantum mechanics.

Seven miles away, Carlos is unlocking his auto-repair garage and plugging in the diagnostic computer he uses to communicate with every car that comes through. The device communicates with the car's processor via a chip containing billions of transistors, each smaller than a virus. Those transistors have no moving parts. They work because electrons inside a thin layer of semiconductor material obey quantum rules that have nothing to do with gears or levers. When his youngest daughter asked what quantum physics is, he surprised himself with his answer: it is the reason your phone works. It is the reason my scanner works. It is everywhere. He was more right than he knew.

Diagram 9.1 - Aisha at the MRI Console

MRI Console

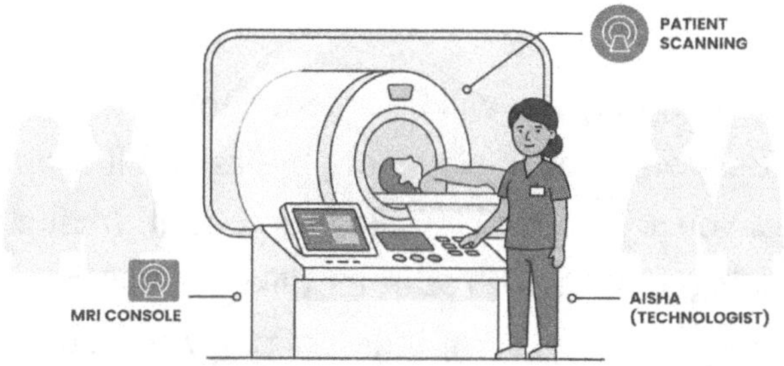

A clean flatt vector MRI technology cr nance a mioodern medical environment

Why It Matters

Look around wherever you are reading this and try to find something that does not depend on quantum physics. A wooden chair? The chemical bonds holding its fibers together are quantum mechanical. The glass in a window? The way silica atoms bond is quantum mechanical. The light from a bulb or the sky? Quantum mechanical, from the electron transitions that produce it to the photons that carry it. You cannot escape quantum effects. Once you see them, they are everywhere.

The technologies in this chapter are not edge cases. They are the backbone of modern civilization. Semiconductors underlie every computer, phone, car, and medical device. Lasers carry internet traffic through glass fibers, read data from discs, and perform eye surgery. MRI machines reveal soft-tissue detail that X-rays cannot show, without exposing patients to radiation. Atomic clocks keep GPS synchronized to within billionths of a second. Quantum computing is just beginning to emerge, promising capabilities no classical computer can match for certain problems.

After this chapter, you will be able to look at your phone and understand it at the level of physics, not just apps and screens. You will know why silicon is special, what a laser is really doing, how an MRI scanner talks to hydrogen atoms, and what a quantum computer is actually computing. That does not make you an engineer. But it makes you something almost as valuable: a person who genuinely understands the world they live in.

Diagram 9.2 - The Quantum Technology Map

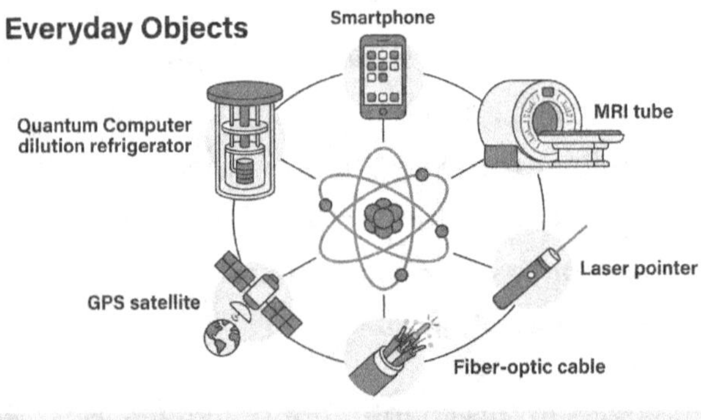

Caption: A man-objective cone tis an flienlly acmsible, an executive tone

The Laser in Your Pocket (Blu-ray, Fiber Optics, and Barcode Scanners)

A laser sounds futuristic even now, decades after it became ordinary. The word stands for Light Amplification by Stimulated Emission of Radiation. To understand it, you need one concept: stimulated emission. And to understand that, you need two things you already know from earlier chapters.

First, atoms absorb energy in specific packets. When a photon of exactly the right energy hits an atom, an electron jumps to a higher energy level. The atom is now excited. Eventually, the electron drops back down, releasing a new photon. This is spontaneous emission. The photon flies off in a random direction. This is ordinary light. A candle, a light bulb, the sun: all of them produce light this way, and all of it is chaotic.

Second, Einstein realized in 1917 that there is another way. If an incoming photon of exactly the right energy passes close to an already-excited atom, it can trigger the atom to emit its stored photon immediately. The new photon emerges in the same direction and with the same energy, perfectly in step with the incoming photon. This is stimulated emission. You now have two photons where there was one, and they are perfect twins.

Diagram 9.3 - Stimulated Emission: One Photon Becomes Two

UNDERSTANDING STIMULATED EMISSION: A KEY LASER PROCESS

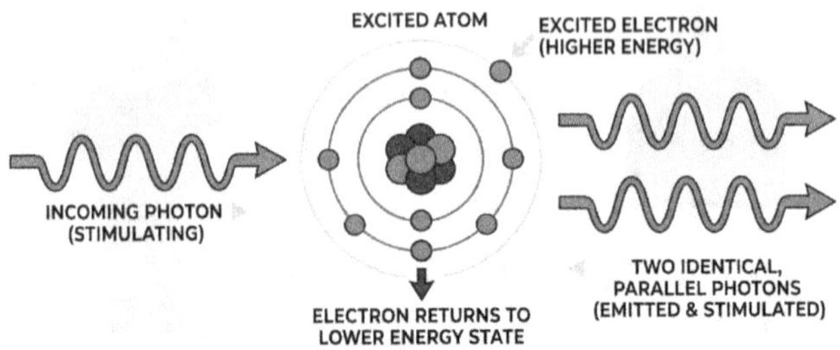

This process produces identical, coherent photons, fundamental for creating laser light.

Now imagine a material packed with atoms energized into excited states, so many that they outnumber the unexcited ones. This is called population inversion. When one photon passes through, it triggers one atom, gains a twin, and those two trigger two more. The number of photons grows explosively, but all of them are identical: same direction, same wavelength, perfect lockstep.

In a real laser, this happens inside a cavity, a tube with a mirror at each end. Photons bounce back and forth, triggering increasingly stimulated emission. One mirror is slightly transparent, so a narrow beam leaks out. That beam is the laser: coherent, monochromatic, and collimated. Those three properties together make it uniquely useful.

Your phone almost certainly contains a tiny laser. A Blu-ray player uses a blue-violet laser because its short wavelength focuses onto much smaller pits on the disc surface, packing in more data than older red-laser DVD systems. A supermarket barcode scanner sweeps a laser across the black-and-white stripes and reads the reflected pattern. These devices all depend on coherence, the property that makes laser light different from the light of a bulb.

Diagram 9.4 - Fiber-Optic Cable with Photons Traveling

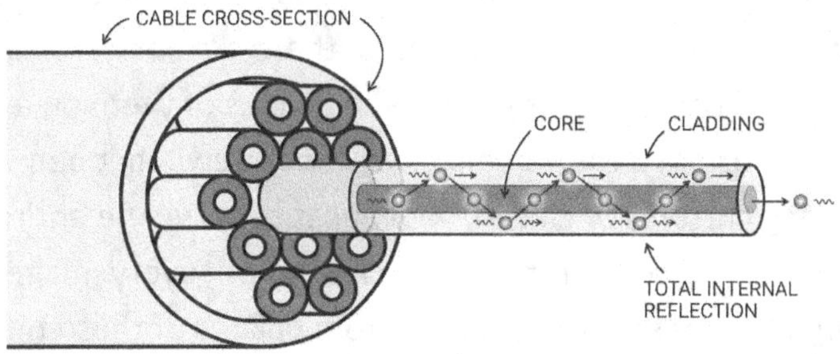

Miniliasline fiber-optic cable

Fiber-optic cables carry most of the world's internet traffic using laser light. A thin strand of glass carries dozens of simultaneous laser beams, each at a slightly different color, each encoding a stream of data. The signal bounces along the inside of the glass by total internal reflection, traveling thousands of miles with minimal loss. The internet, as a physical object, is largely a global network of glass threads filled with laser light.

Lasers also serve medicine. A surgeon can cut tissue with extraordinary precision because the beam focuses to a spot smaller than a cell. Ophthalmologists use lasers to reshape the cornea in LASIK surgery. Stimulated emission, first predicted by Einstein over a century ago, is saving eyesight today.

The Transistor: A Quantum Switch Inside Every Chip

If the laser is the most visible quantum technology in your life, the transistor is the most numerous. There are more transistors in the world today than there are ants. A single modern processor contains tens of billions of them on a chip the size of a thumbnail. Each one is a switch: open, blocking current, or closed, allowing it to flow. The entire complexity of computing, every calculation, every image, every piece of software, is built from billions of these tiny switches clicking billions of times per second.

But a transistor a few nanometers across does not work like a mechanical switch. At that scale, there are no levers to flip, no contacts to touch. What makes it work is the quantum behavior of electrons inside a material called a semiconductor. The key concept is the band gap.

In a conductor like copper, electrons move freely: apply a voltage and current flows. In an insulator like rubber, electrons are tightly bound to their atoms and do not flow. A semiconductor sits between these extremes. Silicon is the classic example. At room temperature, or when a small voltage is applied, electrons can jump across what physicists call the band gap: the energy difference between the locked state and the free-moving state. The size of this gap is a quantum-mechanical property, determined by how electron orbitals overlap in the crystal.

Diagram 9.5 - The Transistor Chip: Quantum Switching

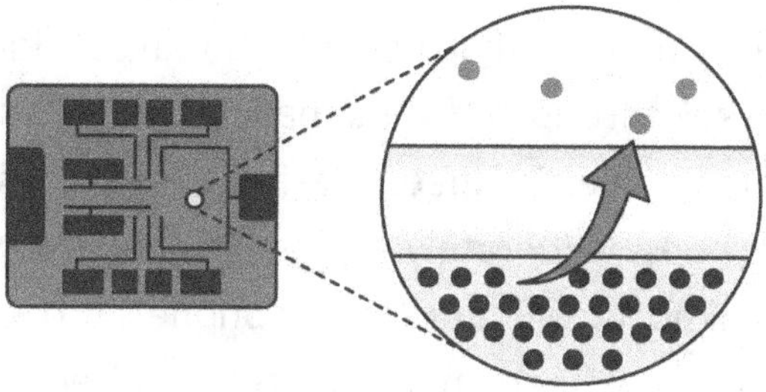

Here is the beautiful part. You can engineer the band gap. By adding tiny amounts of other atoms to silicon, a process called doping, you create regions that are electron-rich and regions that are electron-hungry. Sandwich the two types together and apply a small control voltage; the current can be switched on or off with precision. That is a transistor. The quantum mechanics of the band gap is what makes it work.

Before transistors, in 1947, computers used vacuum tubes: large, hot, fragile, and power-hungry. A single computer filled a room and consumed the electricity of a small neighborhood. The transistor made it possible to shrink that machine onto a desk, then into a pocket, then onto a chip smaller than a postage stamp. Each decade since, engineers have made transistors smaller and faster. The smartphone in your pocket is more powerful than all the computers that guided the Apollo moon missions combined. That is not an exaggeration. It is a direct consequence of shrinking quantum switches.

At today's scales, transistors operate right at the edge where quantum tunneling becomes a real consideration. Electrons leak through barriers that classical physics would say are impassable. Engineers design around these effects, and sometimes deliberately exploit them. The devices in Carlos's diagnostic computer, your phone's processor, and the satellites overhead are not merely inspired by quantum mechanics. They are held together by it.

LEDs, OLED Screens, and the Light of the Modern World

You are almost certainly reading this on a screen that uses quantum mechanics to produce every pixel. Whether you have an LCD panel or an OLED screen, the light it creates comes from electrons making quantum jumps between energy levels. The physics is similar to that of a laser, but without its coherence or directionality. What you get is a controlled, efficient, tunable light source that has quietly replaced almost every older illumination technology.

An LED, or light-emitting diode, produces light when current flows through it. Electrons fall from an electron-rich region to an electron-hungry one, dropping to a lower energy level; the energy they release is emitted as a photon. The color of that photon depends entirely on the size of the band gap. Engineer the material, and you engineer the color of the light. Red LEDs use one semiconductor mixture, green uses another, and blue (the hardest to make, earning its inventors a Nobel Prize in 2014) uses a different compound. Combine all three, and you produce any color in the visible spectrum. That is why your phone screen can display any color.

OLED technology, which stands for organic light-emitting diode, takes this further. Instead of inorganic crystals, OLED displays use thin films of organic molecules that emit light when energized. Each pixel contains red, green, and blue subpixels, each of which can be completely switched off. When a pixel is black on an OLED screen, it is simply off. No backlight bleeds through. This is why OLED screens have such deep blacks and use less power when showing dark images.

Diagram 9.6 - OLED Pixel Grid Glowing

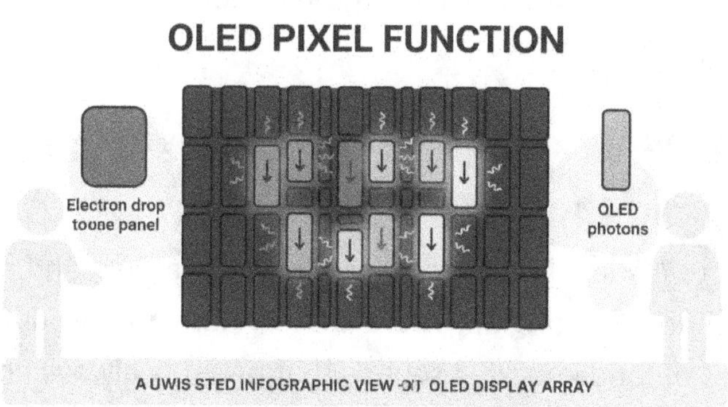

The LED has also transformed general lighting. Traditional incandescent bulbs heat a wire until it glows, and roughly 90% of the energy is lost as heat rather than light. LED bulbs convert electricity directly into light through quantum jumps, with far greater efficiency. An LED bulb lasts twenty times longer than an incandescent bulb and uses a fraction of the electricity. The global shift to LED lighting has saved hundreds of billions of dollars in energy costs and significantly reduced carbon emissions.

Priya, the graphic design student from our cast, works on an OLED laptop screen and photographs her designs on an OLED phone. Every color she sees, every gradient, every bright highlight and deep shadow, is produced by quantum jumps in organic molecules a few nanometers thick. When she exports her designs, the data travels through semiconductor circuits built from quantum switches. When she posts her work online, the signal travels as laser light through fiber-optic glass. The creative tools of the twenty-first century are, from top to bottom, quantum machines.

How MRI Actually Works (Nuclear Spins and the Body as a Radio)

Back to Aisha. The patient is in the MRI scanner, and the machine is starting up. There is a low hum, followed by a series of loud knocking sounds that patients always find surprising. Those sounds come from large electromagnets shifting the magnetic field in precise patterns. But the heart of the machine, the part actually interacting with the patient's body, is doing something much quieter and much stranger.

The human body is mostly water, and water is made of hydrogen and oxygen. A hydrogen atom is the simplest atom: one proton, one electron. That proton has a quantum property called spin. Spin is not literally a rotation, the way a top spins. But it causes the proton to behave like a tiny magnet, with a north pole and a south pole. Normally, the hydrogen protons in your body are oriented in random directions, and their effects cancel out. But inside the MRI scanner's powerful magnetic field, they tend to align, much like a compass needle aligns with Earth's magnetic field. The ensemble of protons develops a net magnetization aligned with the field.

Now the machine does something clever. It sends a pulse of radio waves at exactly the right frequency to match the natural wobble rate of hydrogen protons in that specific magnetic field. This is resonance, and it is a quantum resonance. The protons absorb radio energy, and their spins tip away from the direction of the field. When the radio pulse ends, the protons relax back to equilibrium. As they do, they emit their own burst of radio waves. The scanner's antenna picks up that emission.

Diagram 9.7 - MRI: Hydrogen Spins in a Magnetic Field

MRI Tube Scanner

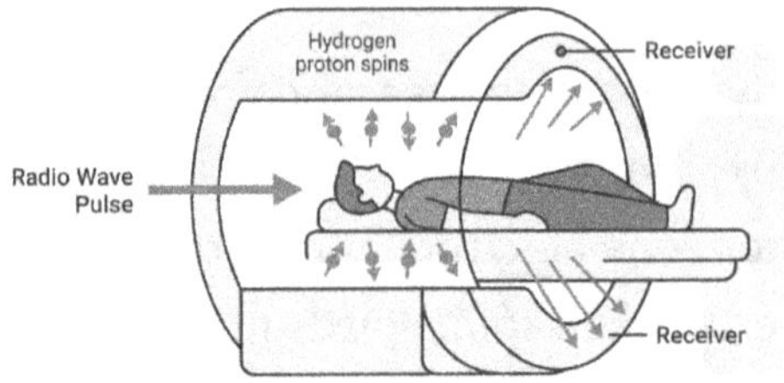

The svarn hydrogen proton spins are aligning tw the strong magnetic field of the MRI.

Different tissues contain hydrogen in different chemical environments. Fat, muscle, bone, and tumor tissue each cause hydrogen protons to relax at slightly different rates after the pulse. The scanner measures those rates and uses them to build contrast. A computer turns the pattern of radio emissions into a three-dimensional image that can distinguish a healthy kidney from a tumor, intact muscle from a torn ligament, and a healthy brain from a multiple sclerosis lesion. All from listening to hydrogen atoms relax.

There is no radiation involved. X-rays carry enough energy to damage DNA, which is why exposure is limited. MRI uses radio waves, far lower in energy and harmless to tissue. This is why MRI has become so important: it reveals soft-tissue detail that X-rays cannot show, with no radiation risk. Functional MRI can also track blood flow, allowing researchers to watch which parts of the brain respond to different experiences by measuring the quantum magnetic properties of iron atoms in hemoglobin.

The machine Aisha is operating will take about thirty minutes. It pulses the magnetic field hundreds of times, collecting signals from slightly different positions, then assembles them into a detailed map of the patient's abdomen. The quantum mechanics of nuclear spin is serving medicine in the most direct possible way.

GPS, Atomic Clocks, and the Precision of Time

The next time your phone tells you that you are standing at a specific address, appreciate how improbable that is. A satellite orbiting twenty thousand kilometers above you, moving at fourteen thousand kilometers per hour, can locate you within a few meters. The technology that makes this possible centers on a quantum device: the atomic clock.

GPS works by triangulation. Your phone receives signals from multiple satellites, each stamped with the exact time it was sent. The phone measures how long each signal took to arrive and calculates distances. The keyword is precision. Light travels at 300,000 kilometers per second, so if satellite timestamps are off by just one millionth of a second, the calculated position is wrong by 300 meters. For meter-level accuracy, the clocks must be right to within billionths of a second. Atomic clocks can do this. Ordinary mechanical clocks cannot come close.

Diagram 9.8 - GPS Satellite with Atomic-Clock Icon

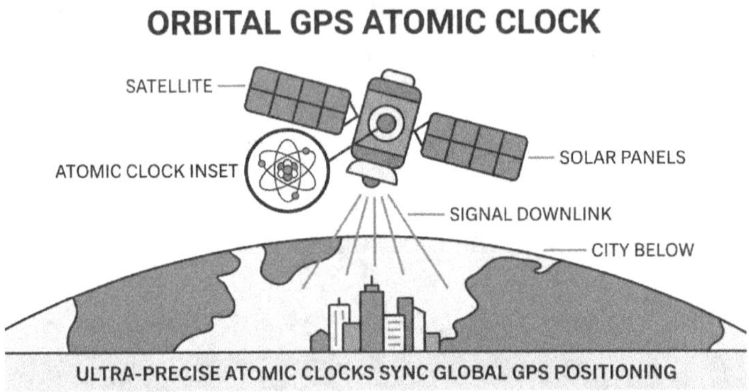

An atomic clock uses the quantum properties of atoms to keep time. Atoms have specific, unchanging natural frequencies set by quantum mechanics. They do not change with temperature, pressure, or age. The most common type uses cesium atoms. A beam of cesium atoms passes through a microwave cavity tuned to match a specific quantum energy transition in the atom. The clock counts those transitions. Modern cesium clocks are accurate to one second in three hundred million years.

There is also a relativity correction built into GPS. The satellites move fast relative to the ground, so their clocks tick slightly more slowly according to special relativity. They are also higher in Earth's gravitational field, so they tick slightly faster due to general relativity. The two effects do not cancel, and the net result is that satellite clocks run about thirty-eight microseconds per day faster than ground clocks. At the speed of light, that error accumulates to eleven kilometers per day. Every GPS receiver applies this relativistic correction. Your phone does general relativity every time it tells you to turn left.

Atomic clocks also underlie financial transactions and telecommunications networks: systems that require precise time synchronization across geographically separated locations. The quantum mechanics of cesium energy levels is, quietly, part of what keeps global finance running.

The Computer That Thinks in Superpositions

Everything we have discussed so far uses quantum properties as a background ingredient in devices that still operate classically overall. A transistor exploits band gaps, but once it is on or off, it behaves classically. An MRI machine exploits quantum spin, but the image it produces is a classical picture. The next generation is fundamentally different. It operates on quantum states directly. It computes using superposition and entanglement as active resources. That technology is the quantum computer.

An ordinary computer processes information in bits: zero or one. A processor with ten billion transistors performs billions of operations per second. This is extraordinarily powerful. Yet for certain types of problems, even the fastest classical computer would take longer than the age of the universe to find a solution.

A quantum computer processes information in qubits, short for quantum bits. A qubit can be zero or one, like a classical bit. But it can also be in superposition: both zero and one simultaneously, like the spinning coin from

Chapter 5. This sounds like a curiosity until you think about what happens with many qubits.

With two qubits in superposition, the system simultaneously represents all four combinations: 00, 01, 10, and 11. With ten qubits, all one thousand and twenty-four. With three hundred qubits, more combinations than there are atoms in the observable universe. This exponential growth is the source of quantum computing's potential power.

Diagram 9.9 - Qubits as Spinning Coins in an Array

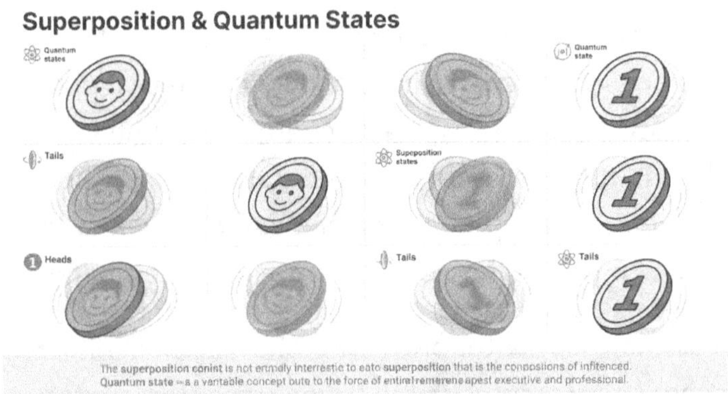

Holding many states simultaneously is not enough. You also have to do something useful with them. A well-designed quantum algorithm arranges the computation so that states corresponding to wrong answers interfere and cancel out. In contrast, those corresponding to right answers interfere constructively and reinforce each other. When you measure the qubits, the right answer is the one that survives. This use of interference to extract useful information is the central trick of quantum computing.

Think of it this way. Imagine a maze with a million paths and only one exit. A classical computer tries paths one at a time, very quickly. A quantum computer explores all paths simultaneously and uses interference to amplify the path that leads out while suppressing the dead ends. This is not quite a precise technical description of every quantum algorithm. Still, it captures the spirit: the power comes from exploring many possibilities at once and using quantum physics to extract the answer.

Diagram 9.10 - The Gold Dilution Refrigerator Quantum Computer

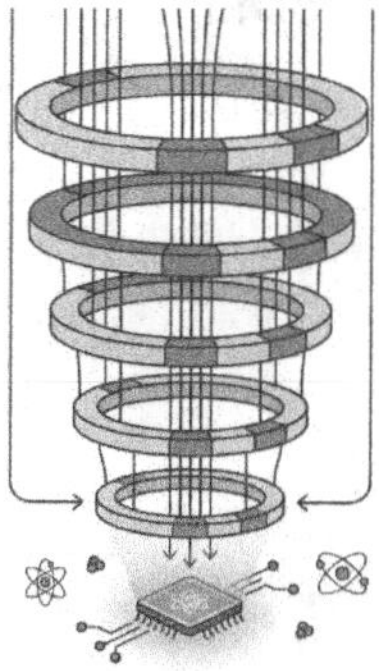

dilution refrigeerator quantum computer

Building a quantum computer is enormously difficult. The central challenge is decoherence: any interaction between a qubit and the outside world prematurely collapses its superposition into a definite state, thereby causing the computation to fail. Quantum computers must be cooled to temperatures close to absolute zero, far colder than outer space, to eliminate thermal noise. The gold cylindrical structure in photographs of quantum computers is a dilution refrigerator that cools the qubit chip to about fifteen millikelvin, a hundredth of a degree above absolute zero.

What Quantum Computers Will and Will NOT Do (No Hype)

Quantum computers attract breathless coverage. They will break all encryption. They will solve every problem instantly. They will make classical computers obsolete. None of these claims is accurate. Quantum computers are real; they are powerful for specific applications, and they will eventually be transformative. But they are not magic, and they are not a replacement for classical computing.

Diagram 9.11 - Quantum Hype vs. Reality

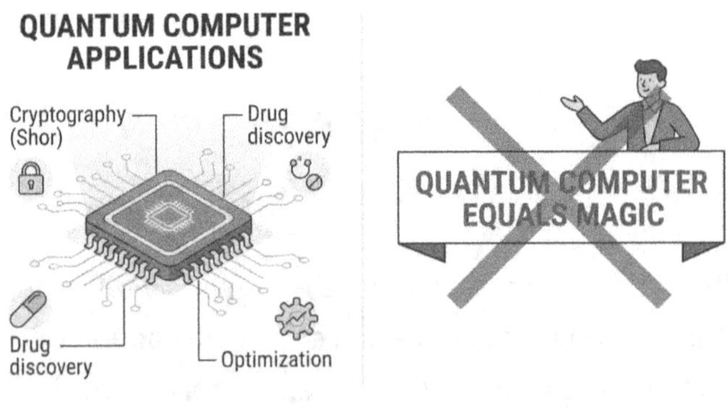

Quantum computers are well-suited to problems whose structure maps onto quantum interference, enabling dramatic speedups. The most famous example is factoring large numbers, the mathematical problem underlying most current encryption. A classical computer must try enormous numbers of possibilities to find the prime factors of a large number. A quantum algorithm discovered by Peter Shor in 1994 can solve this problem exponentially faster. If a quantum computer with enough high-quality qubits were to run Shor's algorithm, it could break most current encryption. This is a serious concern, and mathematicians are already developing new encryption methods resistant to quantum attacks.

Simulating quantum systems is another area of genuine promise. Classical computers are poor at modeling electron behavior in molecules because quantum interactions become exponentially more complex as the number of electrons increases. A quantum computer, which operates natively according to quantum rules, can simulate these interactions more directly. This could transform drug discovery: designing a new drug molecule is fundamentally a quantum chemistry problem. It could also transform materials science, allowing engineers to design new superconductors and battery materials by simulating their quantum properties before synthesizing them.

Here is what quantum computers are not good at: general-purpose computation. They cannot run your word processor faster, stream video, or play games. For the vast majority of tasks, classical computers are far more efficient. A quantum computer performing a classical calculation is like using a concert grand piano to play a single note. The future is not quantum computers replacing classical ones. It is quantum computers solving the specific hard problems that classical computers cannot, while classical computers handle everything else.

Large-scale, fault-tolerant quantum computers do not yet exist. Current machines are NISQ devices, Noisy Intermediate-Scale Quantum systems. They have demonstrated quantum advantage on carefully chosen benchmark tasks, but those tasks were designed to showcase hardware rather than solve urgent practical problems. The transition to millions of low-error qubits with reliable error correction may take decades. Anyone who says quantum computers will transform everything by next year is overstating the case. Anyone who says they will never be practical is wrong, too. The truth is in between, and it is genuinely exciting.

The Quantum World in Living Things

Before turning to what all of this means for your daily life, there is one more frontier worth knowing about: quantum biology. The quantum effects we have discussed so far- band gaps, stimulated emission, and nuclear spin- are phenomena that engineers deliberately harnessed to build machines. But there is growing evidence that evolution discovered some of the same tricks billions of years earlier.

Photosynthesis converts sunlight into chemical energy and is one of the most important reactions on Earth. For decades, biologists assumed it worked through ordinary sequential chemistry: a photon hits a chlorophyll molecule, an electron is knocked loose, and it bumps along a chain of molecules until its energy is captured. But experiments in the early 2000s revealed something unexpected. The efficiency is remarkably high. When researchers used ultrafast laser pulses to observe photosynthesis in real time, they found that energy did not travel in a straight line. It appeared to be exploring all possible paths simultaneously, using quantum interference to find the shortest, most efficient route. Nature had invented quantum optimization long before we thought of the idea.

A second example comes from European robins, migratory birds that navigate across thousands of kilometers with striking precision. Researchers have found evidence that they orient using a compass based on quantum entanglement in proteins in their eyes. A photon from sunlight triggers the creation of a pair of entangled electrons in a molecule called cryptochrome. The entangled pair is sensitive to Earth's magnetic field, and the resulting chemical signal tells the bird which way is up. The bird is running a quantum measurement experiment in its retina every time it navigates.

Quantum biology remains an active research field, and not all researchers agree on the extent to which quantum coherence is central to these processes. But the evidence is strong enough that the field has become serious science. It raises a fascinating question: if evolution discovered quantum coherence as a strategy for efficiency, how fundamental are quantum effects in the fabric of life itself? The quantum world may not just be a layer of physics that makes our technology work. It may be part of what living things are.

Diagram 9.12 - Quantum Biology: Photosynthesis and Bird Navigation

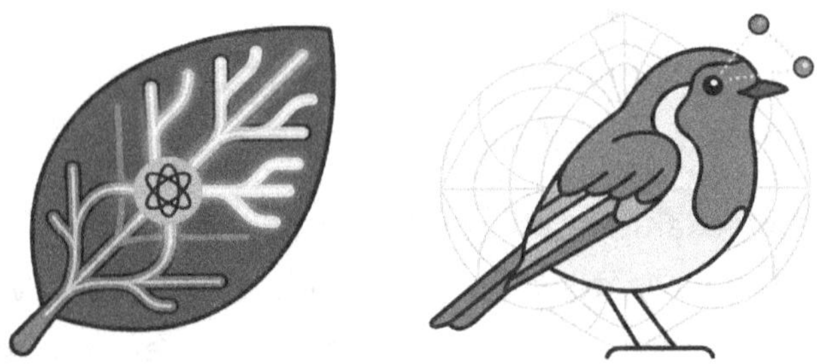

What This Means For You

You now live in a world that looks different from the one you inhabited at the start of this book. Not because the world changed. It was always this way. But because now you can see the layer underneath.

The next time you turn on a light, you know that the photons come from electrons making quantum jumps between energy levels, releasing precisely the energy for that specific color. If it is an LED bulb, the color was engineered by choosing materials with the right band gap. If it is sunlight through a window, those photons were produced in the Sun's core by nuclear fusion. Light is never just light. It is a stream of photons, each carrying exactly the right energy for its wavelength, emitted by electrons according to quantum rules.

The next time you use your phone, you know that the screen is powered by OLED pixels, each one an organic molecule making a quantum jump. The processor is a network of quantum switches, with billions of transistors that use band-gap physics to perform logic. The GPS is synchronized to atomic clocks that measure quantum transitions in cesium atoms, and corrected for relativistic effects predicted by Einstein a century ago. Your phone is not just a gadget. It is a layered anthology of quantum discoveries, assembled into a volume that fits in your hand.

The next time someone in your family needs an MRI, you can explain what is happening: a strong magnetic field aligns hydrogen protons in the body, radio pulses tip those spins, and the machine listens to the protons relax. A computer assembles those signals into an image that can show the difference between healthy and diseased tissue. It is quantum mechanics serving medicine in the most direct way possible.

The next time you hear about quantum computers in the news, you can evaluate what you are reading. A quantum computer uses superposition and entanglement to explore many computational paths simultaneously, then uses interference to extract the useful answer. Current quantum computers are real but limited. The most transformative applications are still being developed. You can tell the difference between genuine progress and hype. That is not a small thing.

And if you are the kind of person who, like Frank from our cast, asks hard questions and will not let anyone get away with hand-waving, you now have the vocabulary to do that. When someone mentions quantum supremacy, you can ask: What exactly did the quantum computer do faster? When someone promises a quantum breakthrough, you can ask: For which specific problem is quantum speedup theoretically guaranteed? These are the questions experts ask. Now you can too.

Takeaway

We started this chapter with Aisha sliding a patient into an MRI scanner and Carlos plugging in a diagnostic computer at his garage. Both of those everyday moments are quantum mechanics made visible, once you know how to look.

Aisha's machine talks to hydrogen atoms in a living body, coaxing their quantum spins into alignment and listening to the radio waves they emit as they relax. Carlos's diagnostic scanner runs on a chip containing billions of quantum switches, each one using the band gap of silicon to control whether current flows. The laser in a fiber-optic cable at the bottom of the ocean produces coherent light through stimulated emission, a process Einstein predicted in 1917. The GPS satellite overhead keeps time using cesium atoms oscillating between quantum energy states. The OLED screen you may be reading this on produces every photon through quantum jumps in organic molecules. The modern world is, in the most literal sense, a quantum machine.

What this chapter wanted you to feel is not just informed but entitled. These technologies are not the property of specialists. They belong to everyone who takes the time to understand them. The quantum mechanics behind a transistor is no more complicated than the quantum mechanics of a photon through two slits. It is the same physics, applied. You have already done the hard conceptual work. You already understand the principles. The technologies in this chapter are those principles put to work.

Quantum computing is the newest frontier, and it deserves both excitement and patience. The same discipline of thinking carefully, distinguishing what we know from what we hope, that made every earlier quantum technology possible, will determine how transformative quantum computing ultimately becomes. The physicists and engineers working on it are building on the same foundations you have been reading about: superposition, entanglement, and interference. They are not working with magic. They are working with physics that is strange enough to seem like magic, once you understand it well enough to appreciate just how strange it really is.

Aisha's patient, it turns out, has a benign cyst, not a tumor. The MRI scan shows it clearly: the fluid inside is bright against the darker kidney tissue. She will deliver that news in about an hour. The relief on the patient's face will be one of those moments that remind her why she became a nurse. The quantum physics she does not think about while she works will have played its quiet, essential role: turning the spin states of hydrogen atoms in a living body into a picture that saves someone from a week of terrifying uncertainty. That is the whole story—physics in the service of people—Quantum mechanics at work.

10 Many Worlds and Open Questions: What Quantum Mechanics Still Cannot Tell Us

Opening Scenario

Theo sits on the wooden bench at the far end of his garden. The sky has gone from orange to deep violet, and the first few stars are becoming visible above the roofline.

The tomatoes have been staked, the tools are hung, and there is nothing left to do tonight except think. He has a cup of chamomile tea balanced on his knee, and he is thinking about a coin.

Back in Chapter Five, Theo learned that a quantum particle does not simply have a hidden value waiting to be discovered. It genuinely holds multiple possibilities at once, the way a coin spinning in the air is neither heads nor tails until it lands. That idea lodged somewhere in his mind and stayed there. He mentioned it to his granddaughter over video call. He brought it up with his neighbor, and the two of them puzzled over it for twenty minutes before agreeing they were both confused. Theo has been reading, listening, and quietly turning the idea over ever since. Tonight, sitting under a sky full of ancient light, he wants to push the question one step further. He does not just want to know what quantum mechanics does. He wants to know what it means.

That question, it turns out, is the one that physicists themselves cannot agree on. The theory is almost a hundred years old. It has been tested to an extraordinary degree of precision. It has given us transistors, lasers, and the entire digital age. And yet some of the smartest people who have ever lived still argue passionately about what it is actually saying about reality. Tonight, sitting with Theo in his garden, we will find out why.

Diagram 10.1 - Theo in the Evening Garden

Theo Garden Infografhit

Theo reflects on the day's events as evening falls, enjoying a peaceful moment in his garden, framed by silstotted plants and the fading light.

Why It Matters

You might wonder why it matters what quantum mechanics means, as long as it works. After all, your phone works whether or not you have thought about wave functions. The GPS in your car gets you to the right street, whether or not you understand superposition. Why not just accept the results and move on?

That is a reasonable position, and it is the one most working physicists take on a Tuesday afternoon when they are in the middle of a calculation. But there are moments when the interpretive question stops being philosophical and becomes practical. If you want to build a quantum computer, you need to know whether quantum states are real physical things or just bookkeeping tools. If you want to understand how quantum effects relate to gravity, you need a framework for what quantum mechanics is actually describing. The meaning is not just a luxury. It is the signpost for where science goes next.

There is also a simpler reason to care. We are the species that looks up at the stars and asks what they are. Quantum mechanics is the deepest description of reality that science has ever produced, and it leaves an unresolved question at its center. That question is the measurement problem. Understanding it, along with the three main attempts to answer it, is what this chapter is about.

By the time you finish reading, you will be able to explain what Copenhagen actually claims, what Many-Worlds proposes, and what pilot-wave theory offers. You will understand decoherence and why it explains so much without settling the deepest question. You will know what sits at the frontier: quantum gravity, dark matter, and the still-unresolved relationship between the quantum world and the one we experience. And you will leave with something Theo himself will carry out of his garden tonight: not unease, but the particular satisfaction of standing at the edge of something vast and genuinely thrilling.

Diagram 10.2 - Three Physicists at a Chalkboard

INFOGRAPHC

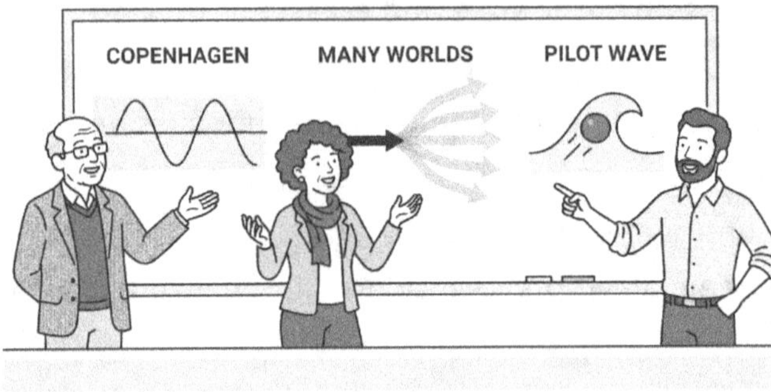

What Everyone Agrees On

Before we reach the disagreements, it is worth spending a moment on the remarkable agreement. The equations of quantum mechanics are not in dispute. Every serious physicist, regardless of their interpretation preferences, uses the same mathematics to make predictions. Those predictions have been tested more rigorously in experiments than almost any other idea in the history of science.

A wave function can describe any quantum system, a mathematical object that encodes the probabilities of different outcomes. If you measure the spin of an electron, the wave function tells you the chances of finding it spinning one way versus another. Between measurements, the wave function evolves smoothly and predictably. Think of it as the rule governing how a cloud of possibilities shifts over time. It is elegant, precise, and agreed upon by everyone.

Everyone also agrees on superposition. An electron can have its spin genuinely pointing in two directions at once, not just unknown to us, but actually undecided until a measurement forces a choice. The double-slit experiment makes no sense unless each photon is genuinely passing through both slits simultaneously. Entanglement is also agreed upon: two particles can be linked so that measuring one instantly determines something about the other, regardless of the distance between them.

And here is the piece that everyone agrees on but nobody fully understands: when a measurement happens, the spread of possibilities narrows to a single outcome. You get one result. The wave function, which encoded all those probabilities, suddenly collapses into a single definite state. This change is called wave function collapse. Everyone agrees that something like this occurs. They disagree, sharply and passionately, about what it actually is.

What Nobody Agrees On: The Measurement Problem

Here is the puzzle at the heart of quantum mechanics, stated as plainly as it can be. The equation governing the evolution of quantum systems over time is deterministic and continuous. Given the wave function right now, you can compute it exactly at any future moment. Nothing sudden. Nothing jumpy. Just smooth, predictable change.

But measurement breaks that smooth story. When you measure a quantum system, you do not get a spread of possibilities. You get one result. The electron is spinning one way, not the other. The photon arrives at this detector, not that one. Something collapsed from a cloud of probabilities to a single definite outcome. And the standard evolution equation cannot, on its own, explain why that happens. This is the measurement problem. It is not a technical loose end. It is a fundamental gap between the mathematical machinery of the theory and what actually happens in a laboratory.

Theo's spinning coin is a decent analogy. While the coin spins, it genuinely is not heads or tails. That is superposition. When it lands, you get one result. That is like a collapse. But with a coin, there is an obvious physical explanation: air resistance, the angle of the throw, and the surface of the table. With quantum mechanics, the rules that govern the system before measurement are genuinely different from the rules that seem to apply at the moment of measurement, and no one is sure why.

Part of the difficulty is that the word measurement is slippery. In everyday life, measurement means using a ruler or a scale. It sounds passive. But in quantum mechanics, a measurement is an interaction between the quantum system and something large enough to produce a definite result: a detector, a photographic plate, a gas molecule. At what point does the spread of possibilities become a single definite fact? Where is the line between quantum and classical? Nobody has drawn that line convincingly.

Different physicists answer the measurement problem differently. Some say the wave function is just a calculation tool, so the question of what collapses it is meaningless. Some say the universe branches at every measurement, so no collapse occurs at all. Some say the wave function represents a real physical wave that guides the particle, and that apparent collapse is just our discovery of where the particle was all along. These are the three main interpretations.

Diagram 10.3 - Wave Function Cloud Collapsing to a Point

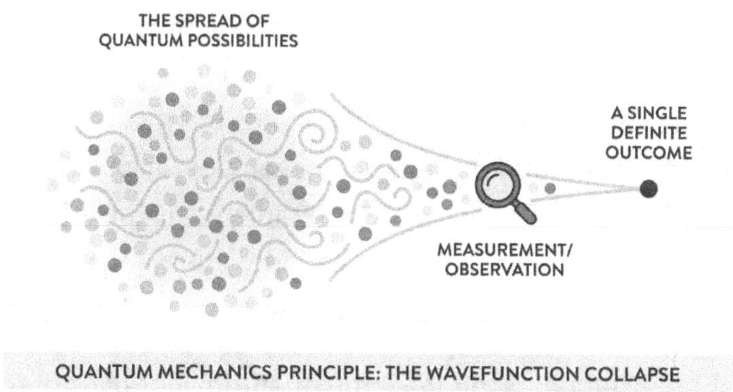

Copenhagen: Shut Up and Calculate

The oldest and most widely held interpretation is called the Copenhagen interpretation. It takes its name from the city where Niels Bohr worked, and was developed primarily by Bohr and Werner Heisenberg in the late

1920s. It is a pragmatist's answer to a philosopher's question.

The Copenhagen view says this: the wave function is a tool for calculating probabilities, not a picture of reality. When we say a particle is in superposition, we are not saying it genuinely exists in two states at once in some deep physical sense. We are saying that our best pre-measurement description of it is a probability distribution. When we measure, we get one result. The wave function updates. We move on. Stop asking what was happening before the measurement, because that question has no answer within the theory.

There is a phrase associated with this attitude: 'shut up and calculate.' It was coined by the physicist David Mermin, who used it satirically to describe an overly dismissive response to deep questions. But it captures something real about the Copenhagen approach. The interpretation is explicitly anti-metaphysical. The job of physics is to predict experimental outcomes, and quantum mechanics does that better than any theory in history.

Bohr himself put it this way: the quantum world is not a world of things that have properties independent of observation. It is a world of phenomena, events that occur within the context of measurement. Asking what an electron is doing when no one is looking is not a deep question with a hidden answer. It is an ill-formed question.

But there are real problems with Copenhagen. The biggest one is the division it draws between the quantum and classical worlds. The interpretation treats the measuring device as something fundamentally different from the quantum system being measured. The device obeys classical rules. The system obeys quantum rules. But measuring devices are made of atoms. Atoms are quantum objects. So where is the line? The maps-versus-territories analogy is useful here. The Copenhagen interpretation says the wave function is a map, not the territory. That is defensible. But all of our other great theories, from Newton's gravity to Einstein's relativity, are trying to describe the territory. For many physicists, a theory that refuses to say what reality is actually like has stopped halfway.

Diagram 10.4 - The Map Is Not the Territory

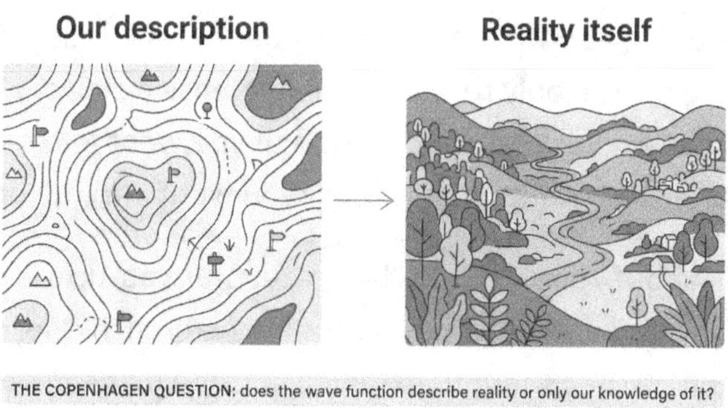

For Theo, sitting in his garden, the Copenhagen interpretation feels a little like being told not to ask what his garden looks like at night when he is not in it. It keeps the theory working, but it leaves a door firmly closed. Not everyone is comfortable with that closed door.

Many Worlds: Every Possibility Comes True Somewhere

In 1957, a young American physicist named Hugh Everett III proposed a radically different answer to the measurement problem. His idea was simple to state and staggering to absorb. What if the wave function never collapses? What if, when a measurement occurs and a quantum system has two possible outcomes, both outcomes happen in two different branches of reality?

This is the Many-Worlds interpretation. Its central claim is that the equation governing quantum evolution is always right, without exception. Superpositions do not collapse. Instead, when a quantum event has multiple possible outcomes, the universe branches. Each branch contains one outcome. Both branches are equally real. Neither is privileged. The only reason you experience a single definite result is that your version of you went into one of the branches. Another version of you went into the other.

Think about Theo's coin. When a quantum particle is measured, both outcomes occur. In one branch, the version of Theo there thinks, well, it came up heads. In another branch, the opposite occurred. Both branches proceed forward. Both are entirely real. The two Theos will never interact or become aware of each other because the decoherence process physically isolates the branches, as we will discuss shortly.

Everett was not writing science fiction. He was trying to take the mathematics of quantum mechanics seriously and see where it led without adding any extra assumptions. His conclusion: the wave function is a real physical object, and when it describes multiple outcomes simultaneously, all of those outcomes actually happen in separate, equally real branches of the universe. The Many-Worlds interpretation has gained serious traction in recent decades. Its advocates include David Deutsch, who coined the term "quantum computer," and physicist Sean Carroll. They argue it is the only interpretation that treats quantum mechanics as a complete, coherent theory without adding anything extra: no collapse mechanism, no hidden variables, just the wave function, evolving according to its equation, all the time.

Diagram 10.5 - Timeline Arrow Splitting Into Parallel Branches

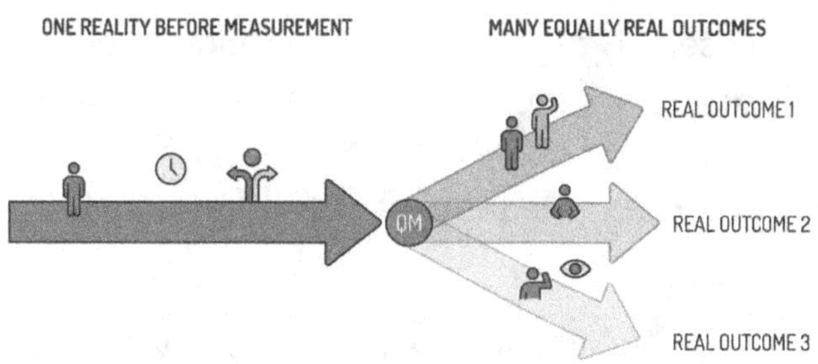

VISUALIZING THE MANY-WORLDS INTERPRETATION OF QUANTUM MECHANICS: ONE TIMELINE, MULTIPLE REALITIES

But Many-Worlds comes with a price: ontological extravagance. If every quantum event branches the universe, and quantum events happen constantly everywhere, then the number of branches is essentially infinite and growing at every moment. Every particle interaction, every photon absorbed by an atom, every radioactive decay creates new branches. This is not saying there are a few parallel universes. It is saying there are unimaginably many, and more are being created right now.

Many Worlds also raises difficult questions about probability. Both outcomes always happen, so where do the probabilities come from? Why are some branches more typical than others? This is an active area of research. Defenders have proposed detailed answers, but critics argue that those answers involve hidden assumptions. What is clear is that Many-Worlds takes the strangeness of quantum mechanics completely seriously. It does not try to explain away the superpositions. It says they are all real, and the appearance of a single outcome is simply because each version of us is in a different branch. In that sense, it is the most radical acceptance of what mathematics is saying.

Pilot Waves: There Is A Real Position All Along

There is a third major interpretation that takes a very different approach. It is called pilot wave theory, or Bohmian mechanics, named after the physicist David

Bohm, who developed it in full in 1952. The basic idea was proposed earlier by Louis de Broglie in 1927, but was largely ignored until Bohm revived and extended it. This interpretation offers something neither Copenhagen nor Many-Worlds provides: a story in which the particle always has a definite position.

Here is the central idea. In pilot wave theory, there are two things, not one. First, there is the wave function, a real physical wave spreading through space. Second, there is the particle, a small, definite object that always has a specific position. The wave guides the particle. The particle surfs the wave. Think of a surfer on the ocean. The water wave is real. The surfer is real. The wave determines where the surfer goes. In pilot wave theory, the quantum wave function plays the role of the ocean, and the particle plays the role of the surfer.

Diagram 10.6 - Particle Riding a Pilot Wave

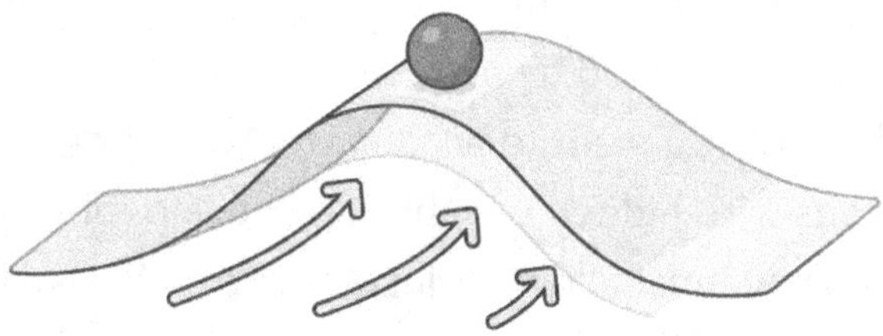

In pilot wave theory, the electron is always somewhere. There is no genuine superposition of position. The wave function spreads out and passes through both slits in the double-slit experiment. But the particle goes through only one slit. The interference pattern builds up because the wave, having passed through both slits, creates preferred landing spots, and the particle, guided by the wave, lands more often in some places than others over many repetitions. This cleanly explains the experiment without any collapse or branching.

Pilot wave theory is deterministic. If you knew the exact position of every particle and the exact wave function of the universe right now, you could calculate every future position of every particle forever. The apparent randomness of quantum mechanics arises because we never actually know those initial positions precisely. The particle's position is a variable that exists but is hidden from us. This is why pilot-wave theories are sometimes called hidden-variable theories.

The most important result regarding hidden variable theories was obtained by physicist John Bell in 1964. Bell proved that if you want to account for quantum predictions with hidden variables, those variables cannot be local. They must allow for instant correlations across any distance. Pilot wave theory is explicitly nonlocal: the wave function is a global object, and changing it in one place affects the guidance of particles everywhere. This is a real feature of the theory, not a flaw, but it means the comfortable classical picture of particles with only local interactions is definitely ruled out.

Critics point out that for many particles, the wave function lives in a space of mind-bending dimensionality. But what pilot wave theory offers that the other interpretations do not is a clear, visualizable story at every moment. The particle always has a position. Measurements reveal it. The randomness reflects our ignorance. Many physicists find this story deeply appealing. Others find the nonlocality and the high-dimensional wave too large a price to pay for definiteness.

Decoherence: Why Big Things Look Classical

There is a concept that helps explain something all three interpretations need to account for: why the quantum world, defined by superpositions and wave-like behavior, gives rise to the ordinary classical world we experience, where objects have definite positions and cats are either alive or dead and not both at once. That concept is decoherence.

Decoherence is not itself an interpretation of quantum mechanics. It is a physical process, described within standard mathematics, that all interpretations agree takes place. Quantum superpositions are fragile. A particle in a superposition can maintain that superposition as long as it is isolated from its environment. But in the real world, particles are rarely isolated. They are constantly interacting with air molecules, photons, and tiny vibrations. Every one of these interactions is a kind of accidental measurement.

Each time a quantum particle interacts with something in its environment, information about the particle's state leaks out. The particle becomes entangled with the environment. Once a quantum system is entangled with a large, complex environment, the superposition effectively disappears from the perspective of any observer in that environment. The different possibilities remain in mathematics, but they no longer interfere with one another. They behave as if they were separate, classical possibilities.

Think of two notes played on a piano at the same time. They interfere, producing a distinctive combined sound. Now imagine those two notes are being played inside a room full of hundreds of other instruments, all making noise. The interference is still there in principle, but it is completely buried. You cannot detect it. For all practical purposes, the notes behave separately. That is decoherence.

Diagram 10.7 - Decoherence: Many States Becoming One Classical Outcome

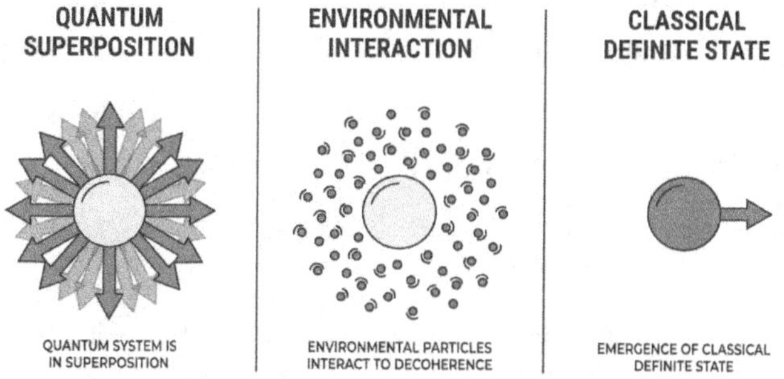

FROM QUANTUM PROBABILITY TO CLASSICAL CERTAINTY: ILLUSTRATING DECOHERENCE AND THE EMERGENCE OF CLASSICAL BEHAVIOR FROM A SUPERPOSITION STATE THROUGH INTERACTION WITH THE ENVIRONMENT.

Decoherence explains why quantum computers are so difficult to build. A quantum computer needs its quantum bits to maintain superpositions long enough to do useful calculations. Any interaction with the environment- a stray photon, a vibration, or a tiny fluctuation in temperature can destroy the superposition. Quantum engineers often cool their systems to temperatures colder than outer space, because even thermal vibrations are part of the environment.

Decoherence also explains why large objects do not appear to be in superposition. A basketball, a planet, and a human being are all composed of enormous numbers of particles, each constantly interacting with the environment. The classical world emerges not because there is a magic line between quantum and classical, but because complexity destroys quantum coherence before it can be detected at our scale. But it is important to understand what decoherence does not explain. It does not solve the measurement problem. Even after decoherence, the mathematics still describes a superposition of outcomes. Decoherence explains why the different outcomes no longer interfere with each other. It does not explain why you experience one and not the other. That gap is still there.

The Open Frontier

Theo looks up at the stars and knows something that Ptolemy did not know, that Newton did not know, and that even Einstein could only partially see: that quantum processes power the stars and that the space between them is threaded with quantum fields and fluctuations. He knows more than any human being has known for most of history. And he is sitting with questions that no human being has yet answered.

The deepest open question in fundamental physics is quantum gravity. Quantum mechanics describes the very small with astonishing precision. General relativity, Einstein's theory of gravity, describes the very large: planets, galaxies, black holes, and the structure of spacetime. Both are extraordinarily successful. And they are fundamentally incompatible. They speak different languages. Quantum mechanics is built around uncertainty and superpositions. General relativity treats space and time as a smooth fabric that bends in response to mass and energy. When physicists try to apply quantum rules to gravity, the calculations produce nonsensical results. At the center of a black hole, and near the moment of the Big Bang, both theories should apply simultaneously. At those scales, we do not have a theory that works.

Diagram 10.8 - Question Mark Constellation in Deep Space

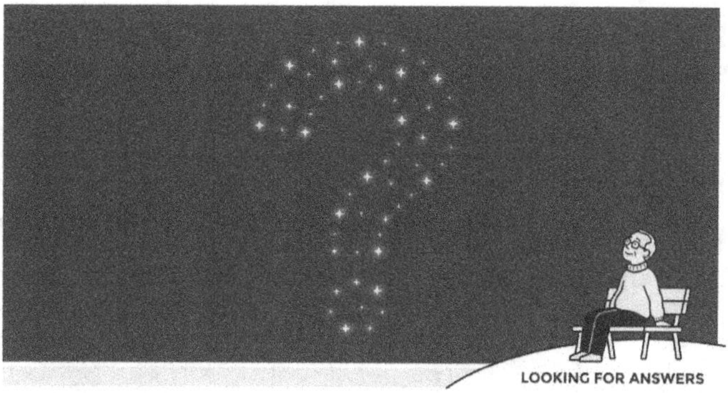

Two main approaches to quantum gravity have been developed. String theory proposes that the fundamental constituents of the universe are not point-like particles but tiny vibrating strings of energy. Different vibration modes correspond to different particles. String theory naturally includes gravity in its mathematical structure. But it requires extra spatial dimensions beyond the three we experience, and it has not yet produced a unique, testable prediction. Loop quantum gravity takes a different approach, attempting to quantize spacetime itself into a network of discrete loops at scales far below anything we can observe. Gravity emerges from the structure of this network. It does not require extra dimensions, but it has not yet produced cleanly distinguishing testable predictions. Both are genuine scientific research programs. The fact that neither has won yet is not a sign of failure. It is a sign of how deep the problem is.

There is also a long-standing question about the relationship between quantum mechanics and consciousness. At what point does the chain of interactions constituting a measurement give way to a definite outcome? Some physicists once proposed that consciousness itself plays a role in collapse. Most modern physicists do not think so, since decoherence provides a physical explanation for why the classical world emerges without invoking minds. But the question has not been fully put to rest.

At the cosmological scale, the pattern of galaxies and voids spanning billions of light-years is thought to have originated in quantum fluctuations in the very early universe. Those tiny quantum variations in density were amplified by the expansion of the universe, resulting in the large-scale structure we observe today. Quantum mechanics left its fingerprints not just on atoms and transistors, but on the entire architecture of the cosmos.

Dark matter is another open frontier. Dark matter is the name for whatever provides the extra gravitational pull that holds galaxies together beyond what visible matter can explain. Several theoretical candidates for quantum particles include hypothetical ultralight particles called axions and supersymmetric particles predicted by extensions of the standard model. Dark matter has not yet been detected directly, and its nature remains one of the most pressing mysteries in physics. The tools used to search for it are quantum-mechanical.

Diagram 10.9 - The Cosmic Web

COSMIC WEB

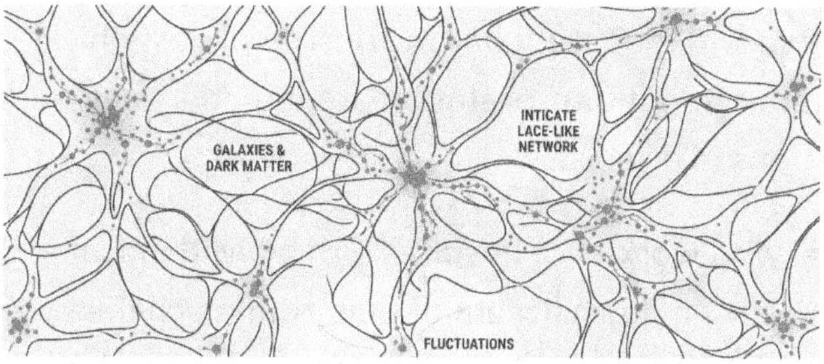

STYLIZED **DEEP-SPACE REPRESENTATION** OF THE COSMIC WEB OF GALAXIES AND DARK MATTER, FORMED BY AN INTRICATE, LACE-LIKE NETWORK, WHERE EARLY UNIVERSE **QUANTUM FLUCTUATIONS** SEEDED THIS VAST ARCHITECTURE.

None of this is settled. All of it is active research. The frontier is real and genuinely open. The questions that remain concern the nature of space and time, the origin of the universe, and the identity of most of the matter in the cosmos. These are the biggest questions there are.

What This Means For You

You might be wondering what any of this has to do with your daily life. The interpretation wars are fascinating, but they do not change how your phone works or what your coffee tastes like. And yet there are several ways these open questions touch your life more than you might expect.

The most immediate is technological. The next generation of quantum computers will depend on a deep understanding of decoherence and how to fight it. The quantum internet, which would use entanglement to create theoretically unbreakable communication channels, is a genuine engineering goal being pursued right now. The pharmaceuticals and materials of the coming decades may be designed using quantum simulations that are only possible with quantum hardware. The philosophical questions about interpretation are, quietly, engineering questions in disguise.

Aisha, who works in a hospital with an MRI machine, is already living downstream of quantum mechanics every working day. Every MRI scan uses the quantum spin of hydrogen nuclei in the body to generate images of soft tissue. The resolution, the contrast, the ability to see a tumor or map a brain without cutting anyone open, all of it rests on quantum principles that were understood only because physicists kept pushing. The open questions of today are tomorrow's technologies.

Maya, who manages software projects, has a stake in quantum cryptography. Current encryption is based on mathematical problems that a sufficiently powerful quantum computer could solve quickly. Governments and technology companies are already working on post-quantum cryptography, new encryption methods resistant to quantum attacks. Understanding that quantum mechanics is real and powerful, and that it won't be replaced by something more comfortable, is practically relevant for anyone who cares about digital security.

Frank, the retired postal worker who always asks the sharpest questions, would put it this way: knowing that even the experts do not agree on what quantum mechanics means makes you a more honest thinker. Science is not a body of settled facts handed down from authority. It is a process of inquiry, argument, and revision. Knowing that changes how you hear scientific claims. You become better at distinguishing what is established from what is contested.

Once you understand that the world, at its deepest level, is genuinely strange, that matter is not little billiard balls. That measurement itself is a physical act with physical consequences; your relationship to what you call the real world quietly shifts. You are living on the surface of a deep ocean. Most of the time, you do not see what is underneath. But knowing it is there is a form of richness that was not available to most human beings throughout history.

Diagram 10.10 - Theo's Thoughtful Face at Dusk

Takeaway

Theo finishes his chamomile tea. The garden is quiet now except for a cricket somewhere near the fence. The stars are fully out.

He has not resolved the measurement problem. Neither has anyone else. He cannot tell you whether the universe branches every time a quantum event occurs, or whether particles always had definite positions guided by invisible waves, or whether the wave function is just a useful fiction. He does not know. Nobody knows. But he knows something tonight that he did not know before he started this book, and it is worth sitting with.

He knows that quantum mechanics is not a gap in our understanding. It is the pinnacle of it. The most precisely tested theory in the history of science, the foundation of every transistor, laser, MRI machine, and quantum computer, is one that its own creators did not agree on. That is not a scandal. That is what deep understanding actually looks like.

He knows that the questions at the frontier- quantum gravity, the possibility of many worlds, and the search for dark matter- are not signs that science is lost. They are signs that science has reached the edge of the map. The edge of the map is always where the most interesting territory begins.

He knows that the vocabulary he has built through this book- superposition, entanglement, wave function, collapse, decoherence, pilot wave, branching, coherence-is not decoration. These are real tools. They let him think more clearly about the world, engage more honestly with physics news, and help his granddaughter understand why her phone works the way it does.

He picks up the empty cup and stands—the bench creaks. The tomatoes are still there in the dark, solid and red and completely quantum mechanical, every atom a tiny storm of waves and probabilities, every chemical bond a product of electrons that settled into their arrangements through the very process this chapter described.

Quantum mechanics cannot yet tell us what its own equations mean about the nature of reality. It cannot reconcile itself with gravity. It cannot explain dark matter. It cannot say whether every outcome happens or just one. These are real limits, honestly held, and they belong to the most successful physical theory ever devised.

What it can do, and what it has done, is give us a picture of reality more accurate, more strange, and more beautiful than anything our ancestors imagined. It tells us that the solid table is mostly space held firm by quantum rules. It tells us that light arrives as individual photons, each following a genuinely uncertain path until it lands. It tells us that the information in a phone exists because quantum tunneling and quantum spin are as reliable as clockwork, even though they arise from probabilities.

Quantum mechanics still cannot tell us everything. But it has already told us more than we had any right to expect. The questions it cannot yet answer are the ones most worth asking.

The cricket goes quiet. Theo goes inside. The stars continue doing what they have always done: burning by quantum fire, sending their light across the dark, arriving at last in the garden of a retired teacher who, tonight, understood a little more of what they are.

Diagram 10.11 - The Cosmic Web and the Quiet Garden

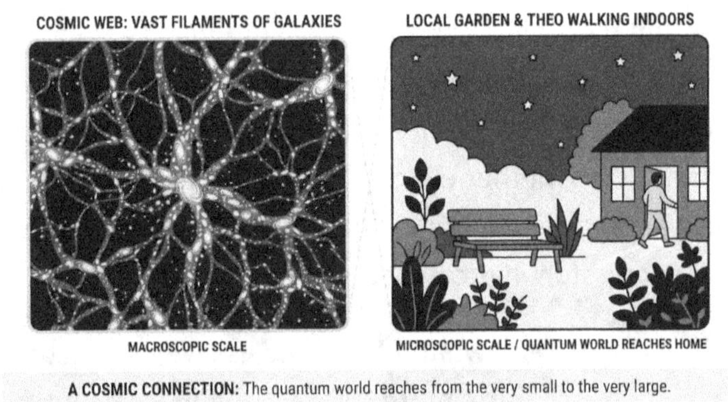

You have reached the end of the journey through quantum mechanics. The world you started in, and the world you now live in, are the same. But you see it differently. That is the gift every good science book tries to give, and it is the one we hope this book has given you.

Diagram 10.12 - Many Arrows Converging on a Single Reader

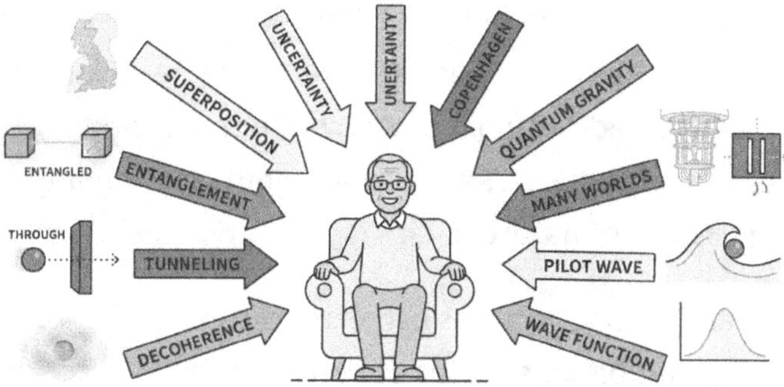

THE READER'S JOURNEY: KNOWLEDGE ACQUIRED AND ASSIMILATED

11 Closing

What We've Walked Through

You started this book in the world of everyday physics. Things fall. Light travels. Machines run hot or cold. That world works beautifully. It lands rockets, builds bridges, and keeps planes in the air. For most of human history, it was the only world anyone needed.

Then came the questions that the world could not answer. Why does hot metal glow red before it glows white? Why does light of the wrong color fail to eject electrons from a metal surface, no matter how bright it shines? Classical physics offered confident answers. Every confident answer was wrong. That was the signal that something deeper was waiting.

Chapter by chapter, you followed the clues. Energy turned out to arrive in packets rather than as a continuous stream. Light turned out to be both a wave and a particle at once. Position and speed turned out to be locked in a permanent trade-off: pin one down precisely and the other blurs. That is not a limitation of measurement tools. It is a feature of nature itself.

You learned that particles can exist in superposition, holding multiple possibilities open at once. You met the wave function, the shape that describes not where a particle is but how likely each possible location is. You watched entanglement link two particles so that measuring one instantly shapes what you find in the other. You saw particles tunnel through barriers they have no business crossing, and you learned that without tunneling, the sun would not shine.

Then you crossed into the world of applications. Transistors, lasers, MRI machines, LEDs. None of these things would exist without the quantum rules that govern the smallest layers of matter. And at the frontier, you met the questions physicists have not yet answered: what exactly happens when a wave function collapses, which interpretation of quantum mechanics is correct, and whether a deeper story is still waiting to be told.

Diagram C.1 - The Reader's Journey

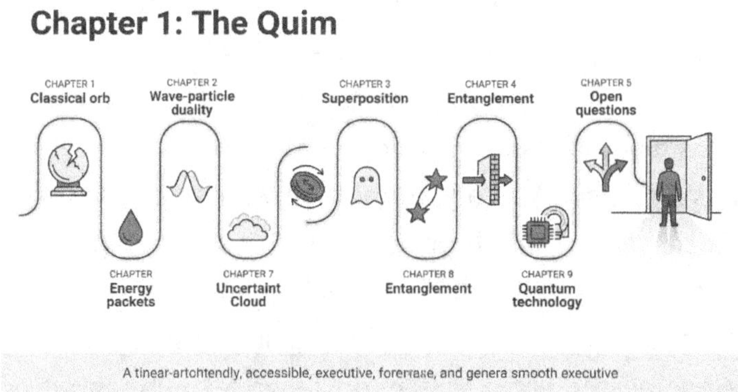

That path is yours. The light coming through the door is not a metaphor for completion. It is a metaphor for a changed way of seeing. The world outside that door is the same world you walked into when you opened this book. But you are not the same reader.

Physics at the everyday scale and physics at the quantum scale are not in conflict. They are nested, the way a neighborhood sits inside a city. You live in the neighborhood. Now you know the city is there and something about how it runs.

What the Cast Carries Forward

Maya checks her phone each morning, knowing that the screen is a semiconductor whose behavior depends entirely on quantum rules, and that the colors she scrolls through are produced by tiny quantum leaps inside LEDs. Theo walks into his garden and notices sunlight differently, aware that it has traveled as a wave and arrived as a particle, and that the fusion keeping the sun burning depends on tunneling. Aisha pushes a patient toward the

MRI scanner. She sees the machine clearly now: magnetic fields flipping quantum spin states of hydrogen atoms, an image drawn by quantum mechanics, not by scalpels. Carlos plugs in his diagnostic tool and knows that every transistor inside it is a gate through which a quantum particle either tunnels or does not. Priya's OLED phone screen produces its colors through quantum leaps in electrons, and she is considering taking a physics course next semester. Frank watched a news segment about quantum computing and leaned forward instead of changing the channel. He understood about half of it. Six months ago, he would have understood none of it.

What We Still Don't Know

Quantum mechanics is the most accurate physical theory ever tested. Its predictions have been confirmed in experiments repeated thousands of times. When the theory says a particle will be found here with this probability, nature agrees.

And yet physicists do not fully agree on what it means. The measurement problem is still open. Something happens when a quantum system is measured: the wave function collapses to a single result. What exactly happens, and why, is not settled. The Copenhagen interpretation holds that the wave function is a tool for prediction, and asking what it is really doing is the wrong question. The many-worlds interpretation says every possible outcome actually occurs in a branching tree of realities. Other interpretations add hidden variables or collapse mechanisms. Each is trying to fill the same gap. Nobody yet knows which interpretation is correct.

There are also the great unifications still ahead. Quantum mechanics and general relativity are both correct in their domains, but do not yet speak the same language. A complete theory of quantum gravity remains one of the deepest open problems in physics. These are not signs that the theory is broken. They are addresses: locations where the next discoveries are waiting.

Diagram C.2 - The Open Frontier

Every open question is an invitation. A century ago, quantum mechanics itself was one of those question marks. The gaps that remain today are in the hands of exactly the right people. Some of the people who will fill them have not yet decided what to study.

A Final Word

You came to this book as someone who wanted to understand. That impulse is the same one that led Planck to look more carefully at glowing metal, that led Einstein to ask what light would look like if you rode alongside it, that led a generation of physicists in the 1920s to sit with results that made no sense until the sense they made was deeper than anything before. Curiosity is not a quality reserved for scientists. It is what brought you here.

You now have vocabulary. You can say wave function and mean something. You can say superposition and not be guessing. You can say "entanglement" and follow the conversation without losing the thread. You understand why the Heisenberg uncertainty principle is not about clumsy measurement but about the texture of reality itself. These are not decorations. They are tools. You can use them.

The world you walk back into is the same world you left. The sunlight. The screen in your pocket. The MRI machine is down the hall. The solar panel on the roof, the LED in the kitchen lamp, the battery in the hybrid at the curb. All the same. But you carry something different now. You carry the knowledge that behind all of it, at a scale too small for any eye to see, the universe holds possibilities open until the last possible moment, lets particles be in two places at once, and connects things across distances that have no business being connected. It has been doing this, quietly, beneath everything, all along.

Look at the sunrise. You know what it is now.

12 References

A–D

Aspect, Alain, et al. "Bell Test Experiments: Closing Loopholes in Quantum Nonlocality." ***Reviews of Modern Physics*****, vol. 96, no. 2, 2024.** (CH7)

Bassi, Angelo, and Sandro Donadi. "Advances in Collapse Models of Quantum Mechanics." ***Nature Physics*****, vol. 20, 2023.** (CH10)

Biercuk, Michael J., et al. "Quantum Control for Next-Generation Quantum Technologies." ***Nature Reviews Physics*****, vol. 5, 2023.** (CH9)

Bravyi, Sergey, et al. "Progress Toward Fault-Tolerant Quantum Computation." ***Nature*****, vol. 624, 2023.** (CH9)

Brown, Harvey R. "Revisiting the Measurement Problem." ***Foundations of Physics*****, vol. 54, 2024.** (CH10)

Cao, Tim, et al. "Ultrafast Electron Dynamics in Solids." ***Science*****, vol. 381, 2023.** (CH3)

Carleo, Giuseppe, et al. "Machine Learning and Quantum Many-Body Physics." ***Reviews of Modern Physics*****, vol. 95, 2023.** (CH6)

Castelvecchi, Davide. "Quantum Sensors Reach New Sensitivity Limits." ***Nature*****, 2024.** (CH4)

Chen, Yu-An, et al. "Quantum Tunneling in Biological Electron Transport." ***PNAS*****, vol. 121, no. 4, 2024.** (CH8)

E–H

Eisert, Jens, et al. "Quantum Entanglement in Modern Experiments." ***Nature Reviews Physics*****, vol. 6, 2024.** (CH7)

Fadel, Matteo, et al. "Macroscopic Quantum Superpositions in Optomechanical Systems." ***Nature Physics*****, vol. 19, 2023.** (CH5)

Fujii, Keisuke, et al. "Advances in Quantum Error Correction." ***Nature Communications*****, vol. 14, 2023.** (CH9)

Gao, Xiaoxing, et al. "Quantum Coherence in Photosynthetic Complexes." ***Science Advances*****, vol. 10, 2024.** (CH9)

Gisin, Nicolas. "Determinism, Nonlocality, and the Quantum World." ***Nature Physics*****, vol. 20, 2024.** (CH7)

Greiner, Markus, et al. "Ultracold Atoms as Quantum Simulators." ***Nature Physics*****, vol. 19, 2023.** (CH6)

Hensen, Bas, et al. "Loophole-Free Bell Tests Revisited with Improved Detectors." ***Physical Review Letters*****, vol. 132, 2024.** (CH7)

Hossenfelder, Sabine, and Tim Palmer. "New Approaches to Quantum Foundations." ***Reports on Progress in Physics*****, vol. 87, 2024.** (CH10)

I–M

Ivanov, Dmitri, et al. "Quantum Tunneling in Next-Generation Semiconductor Devices." ***Nature Electronics*****, vol. 7, 2024.** (CH8)

Juffmann, Thomas, et al. "Matter-Wave Interference with Large Molecules." ***Nature Physics*****, vol. 20, 2024.** (CH3)

Kjaergaard, Morten, et al. "Superconducting Qubits: Current State and Roadmap." ***Annual Review of Condensed Matter Physics*****, vol. 15, 2024.** (CH9)

Kok, Pieter, et al. "Photonics for Quantum Information Processing." ***Nature Photonics*****, vol. 17, 2023.** (CH3)

Lemos, Gabriela, et al. "Imaging with Undetected Photons: New Quantum Optical Techniques." ***Science*****, vol. 382, 2023.** (CH3)

Liu, Zhen, et al. "Quantum-Enhanced MRI Contrast Mechanisms." ***Nature Biomedical Engineering*****, vol. 8, 2024.** (CH9)

Maldacena, Juan. "Quantum Gravity and Spacetime Structure." ***Reviews of Modern Physics*****, vol. 96, 2024.** (CH10)

Matsumoto, Kentaro, et al. "Precision Atomic Clocks and Timekeeping Advances." ***Nature Physics*****, vol. 19, 2023.** (CH9)

N–S

Narang, Prineha, et al. "Quantum Materials for Next-Generation Photonics." *Nature Materials*, vol. 22, 2023. (CH9)

Oreshkov, Ognyan. "Causality and Indefinite Temporal Order in Quantum Theory." *Nature Communications*, vol. 15, 2024. (CH10)

Paternostro, Mauro, et al. "Quantum Thermodynamics: Energy at the Smallest Scales." *Nature Reviews Physics*, vol. 6, 2024. (CH1)

Pusey, Matthew F. "The Reality of the Wave Function Revisited." *Foundations of Physics*, vol. 54, 2024. (CH6)

Riedel, C. Jess, et al. "Decoherence and the Emergence of Classicality." *Reviews of Modern Physics*, vol. 95, 2023. (CH5)

Rosenfeld, Wenjamin, et al. "Long-Distance Entanglement Distribution." *Science*, vol. 381, 2023. (CH7)

Schlosshauer, Maximilian. "Modern Developments in Quantum Decoherence." *Physics Reports*, vol. 1023, 2024. (CH5)

Schumm, Thorsten, et al. "Quantum Sensors for Gravitational Measurements." *Nature Physics*, vol. 20, 2024. (CH4)

T–Z

Torlai, Giacomo, et al. "Neural-Network Quantum States and Wave-Function Reconstruction." *Nature Physics*, vol. 19, 2023. (CH6)

Vool, Uri, and Michel Devoret. "Quantum Electronics: From Josephson Junctions to Qubits." *Nature*, vol. 624, 2023. (CH9)

Wang, Haocun, et al. "Quantum Noise Reduction in Precision Measurement." *Science*, vol. 382, 2023. (CH4)

Zurek, Wojciech H. "Quantum Darwinism and the Emergence of Classical Reality." *Nature Physics*, vol. 20, 2024. (CH10)

www.ingramcontent.com/pod-product-compliance
Lightning Source LLC
LaVergne TN
LVHW022155200926
848481LV00006B/1757

* 9 7 9 8 9 0 4 9 8 0 7 6 4 *